SpringerBriefs in Materials

The SpringerBriefs Series in Materials presents highly relevant, concise monographs on a wide range of topics covering fundamental advances and new applications in the field. Areas of interest include topical information on innovative, structural and functional materials and composites as well as fundamental principles, physical properties, materials theory and design.

Indexed in Scopus (2022).

SpringerBriefs present succinct summaries of cutting-edge research and practical applications across a wide spectrum of fields. Featuring compact volumes of 50 to 125 pages, the series covers a range of content from professional to academic. Typical topics might include

- A timely report of state-of-the art analytical techniques
- A bridge between new research results, as published in journal articles, and a contextual literature review
- A snapshot of a hot or emerging topic
- An in-depth case study or clinical example
- A presentation of core concepts that students must understand in order to make independent contributions

Briefs are characterized by fast, global electronic dissemination, standard publishing contracts, standardized manuscript preparation and formatting guidelines, and expedited production schedules.

Keltoum Khallouq

Exploring High-Temperature Superconductivity in the YBCO System

From Theory to Experiments

 Springer

Keltoum Khallouq
Solid State Physics Laboratory
Faculty of Science Dhar Mehrez
Fez, Morocco

ISSN 2192-1091   ISSN 2192-1105 (electronic)
SpringerBriefs in Materials
ISBN 978-3-031-66237-9   ISBN 978-3-031-66238-6 (eBook)
https://doi.org/10.1007/978-3-031-66238-6

This Springer imprint is published by the registered company Springer Nature Switzerland AG
The registered company address is: Gewerbestrasse 11, 6330 Cham, Switzerland

If disposing of this product, please recycle the paper.

General Introduction

In an era where technology progresses at an unprecedented pace, the quest for materials that can transform our approach to energy, computing, and transportation has become more critical than ever. Among these materials, superconductors stand out as a revolutionary force poised to redefine the limits of technological and scientific innovation.

The study of superconductors is not just an academic pursuit, it is a necessity that holds the key to solving some of the most pressing challenges faced by humanity today. They are utilized in various fields such as medical diagnostics, cancer treatment, fusion reactors, particle accelerators, and ultra-cold atoms physics. These materials—ranging from high-temperature superconductors to superconducting meta-materials—have revolutionized technology by enabling advancements in opto-electronics, magnetic resonance imaging, magnetic energy storage, and even magnetic levitation trains [1–3].

Understanding the history, basic principles, and recent discoveries in superconductivity is essential for harnessing its full potential in diverse applications. The continuous research and development in superconductors promise a future where they will continue to enhance the quality of daily human life and drive innovation in various scientific and technological domains.

Superconductors, materials that can conduct electricity without resistance when cooled to critical temperatures, offer a glimpse into a future with drastically reduced energy losses. This capability alone makes them an invaluable resource in our ongoing battle against climate change.

Today, approximately 26%–30% of generated energy is lost in the current electrical distribution system due to wire resistivity, making the system only 70%–74% efficient [4], [5]. To address these losses, utilizing direct current (DC) transmission instead of alternating current (AC) has been proposed as a more efficient solution, with DC transmission lines costing less per mile and ultimately saving energy and money over their lifespan [6]. Therefore, while 5% energy loss during transmission is a common benchmark, the current conventional systems experience higher losses, highlighting the importance of exploring more efficient transmission technologies like DC to minimize energy wastage [7]. Superconductors can cut this loss close to zero, promising a future of more efficient power grids and sustainable energy systems.

The implications for renewable energy are particularly profound. Integrating superconducting materials into wind turbines and solar panels could significantly enhance their efficiency and output. This would not only make renewable energy more viable and cost-effective but also accelerate our transition from fossil fuels, a critical step in combating global warming.

Moreover, superconductors are at the heart of some of the most advanced experimental technologies in the world. The Large Hadron Collider (LHC), the world's largest and most powerful particle accelerator, relies on superconducting magnets to steer beams of particles at nearly the speed of light. Without superconductors, such experiments, which are crucial for unlocking the mysteries of the universe, would not be feasible.

In the realm of computing, superconductors present an opportunity to leapfrog the limitations of current semiconductor-based technologies. Quantum computers, which promise to revolutionize fields from drug discovery to cryptography, depend heavily on superconducting materials for their qubits, the basic units of quantum information. The development of superconducting materials for quantum computers could vastly increase their reliability and speed, opening up new horizons for computing performance.

Transportation technologies also stand to benefit immensely. Superconducting materials are key to the development of maglev (magnetic levitation) trains, which promise faster, quieter, and more energy-efficient public transport options. In cities around the world, where urban congestion and pollution are growing concerns, maglev trains could provide a cleaner, more sustainable alternative to traditional rail and road transport.

Despite their potential, superconductors face significant challenges, primarily related to their need for extremely low operating temperatures. However, recent advancements in high-temperature superconductivity research have begun to lower these barriers, enhancing the practicality and accessibility of superconductor-based technologies.

The ongoing study of superconductors is, therefore, not merely an academic endeavor but a critical investment in the future of our planet. It holds the promise of energy systems that are both more efficient and more environmentally friendly, of medical technologies that can save more lives, of computers that can solve previously intractable problems, and of transportation systems that could redefine how we travel. Ignoring the potential of superconductors could mean missing out on some of the most significant technological leaps of our time. Hence, the continued exploration and development of these extraordinary materials are not only important but also essential for the advancement of our civilization.

In the realm of condensed matter physics, the discovery of high critical temperature (high-T_c) superconducting oxides has marked a paradigm shift, promising revolutionary applications in technology and profoundly influencing our theoretical understanding of superconductivity.

This book presents a comprehensive exploration of high-T_c superconducting oxides, with a spotlight on the Yttrium Barium Copper Oxide (YBCO) system, a cornerstone in the field of superconductivity.

Chapter 2 delves deep into the world of high critical temperature superconducting oxides of the YBCO system. It begins with an introduction to these marvels of modern physics, presenting the materials that have captivated scientists' and engineers' imaginations alike. The chapter progresses by dissecting the perovskite structure, pivotal to understanding the superconducting properties, and explores the $YBa_2Cu_3O_{7-\delta}$ structure, laying the groundwork for comprehending the intricate behaviors of these materials. Special attention is given to the CuO_2 planes, the heart of superconductivity in these oxides, and expands the discussion to other structures, enriching the reader's grasp of the material's complexity.

In the subsequent sections, a focused study on the YBCO prototype is presented, detailing the crystal structure, the phase diagram, and the significance of the irreversibility line. It also discusses the effects of substitution in $YBaCu_3O_{6+\delta}$, providing insights into the material's flexibility and the impact on its superconducting properties.

Chapter 3 transitions to the experimental battlefield, where the secrets of the YBCO system are unlocked. It guides the reader through the meticulous preparation of samples, the rigorous process of calcination, grinding, compacting, and sintering, and the critical thermal treatments that lead to the formation of superconducting phases. The chapter also covers the sophisticated techniques of X-ray diffraction and iodometry, indispensable tools in the characterization of these materials, and concludes with the methods of measuring alternating magnetic susceptibility and electrical resistivity, key to understanding the superconductors' behavior under different conditions.

As the book unfolds, it ventures into the effects of substitutions and heat treatment on the structural and superconducting properties of the $Y_{0.5}Ln_{0.5}SrBaCu_3O_{6+\delta}$ system, expanding the horizon of knowledge on how subtle changes can influence superconductivity. It presents experimental techniques, discusses results, and engages in a thorough discussion, culminating in a comprehensive conclusion that ties the findings together.

In the concluding chapters, the book takes a broader view, discussing alternating magnetic shielding and resistivity in high critical temperature superconductors, showcasing the multifaceted nature of these materials and their potential to revolutionize various technological domains.

This book is an invitation to journey through the fascinating world of high-T_c superconductors, aimed at researchers, students, and anyone intrigued by the frontier of superconductivity. It presents not just a collection of data and findings but a narrative of discovery, challenge, and the relentless pursuit of understanding the mysteries locked within high critical temperature superconducting oxides.

References

1. S.S. Ali, *Superconductors for Medical Applications* (Materials Research Forum LLC, 2022), pp. 211–229
2. R. Combescot, *Superconductivity: An Introduction* (Cambridge University Press, 2022)
3. Inamuddin, T. Altalhi, V. Gupta, M. Luqman (eds.), *Superconductors—Materials and Applications* (Materials Research Forum LLC, 2022)
4. Jamius Bin Stepanus Herybertus Rivanaldi Podajow, Adelhard Beni Rehiara. Analysis of energy losses in medium voltage network conductors based on load curves at pt. pln (persero) up[3] manokwari area. JISTECH: J. Info.Sci. Tech. **12**(1), 166–174 (2023)
5. A. Akhtar, F. Mahmood, R. Khan. A comprehensive review on wireless power transmission. In *2022 International Conference on Electrical Engineering and Sustainable Technologies (ICEEST)*. IEEE, December 2022
6. R.B. Spielman, S. Chantrenne, D.H. McDaniel. Energy losses in high current density conductors. In *2007 IEEE 34th International Conference on Plasma Science (ICOPS)*. IEEE, June 2007
7. J. Oestergaard, J. Okholm, K. Lomholt, O. Toennesen. Energy losses of superconducting power transmission cables in the grid. IEEE Transact. Appil. Superconduct. **11**(1), 2375–2378 (2001)

Contents

Chapter 1
Generalities and Theoretical Models for YBCO System

1.1 Introduction

1.1.1 Discovery of Superconductivity

Superconductivity was discovered in mercury by the Dutch physicist Heike Kamerlingh Onnes in 1911, at the University of Leiden in the Netherlands, 3 years after he successfully liquefied helium [1–4]. He observed the elimination of electrical resistivity below a temperature of $-268.15\,°C$ (4.2 K) as documented in his experimental notebook (Fig. 1.1). This phenomenon, wherein an electric current can flow without energy loss, indicated a transition from the normal conducting state to the superconducting state at a temperature known as the critical temperature (T_c). A year after this discovery, Onnes observed the same phenomenon in tin and lead, the latter, which is a poor conductor at room temperature, loses its resistance below a temperature of 6 K, and tin becomes superconducting at 3.7 K [5, 6]. For this discovery H. Kamerlingh Onnes awarded the Nobel Prize in 1913 [7], he coined the term "superconductivity" for this new state of electronic properties of matter.

In 1933, W. Meissner and R. Ochsenfeld discovered the perfect diamagnetism associated with the superconducting state [8]. This property, known as the Meissner effect, involves the exclusion of all external magnetic field penetration thanks to the circulation of supercurrents in the superconductor.

In 1935, the London brothers demonstrated that the Meissner effect is a consequence of minimizing the free energy carried by the superconducting current [8, 10, 11]. In 1950, Ginzburg and Landau proposed a phenomenological theory that successfully explained the macroscopic properties of superconductivity using the Schrödinger equation. In the same year, Maxwell and Reynolds et al. observed that T_c varies with the isotopic mass, concluding that lattice vibrations and therefore electron–phonon interactions are crucial factors.

K. Khallouq, *Exploring High-Temperature Superconductivity in the YBCO System*, SpringerBriefs in Materials, https://doi.org/10.1007/978-3-031-66238-6_1

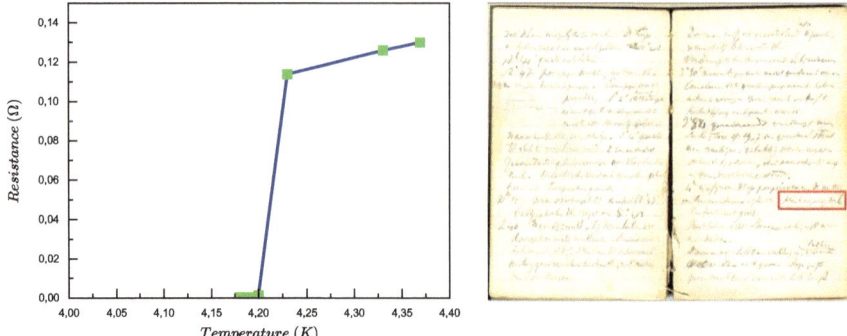

Fig. 1.1   (Right) A copy of Heike Kamerlingh Onnes's experimental notebook. On the right page, we can read "Kwik nagenoeg nul" which means "Mercury practically zero". He was, of course, referring to its electrical resistivity. (Left) Electrical resistivity as a function of temperature [9]

In 1957, A. Abrikosov discovered that based on a phenomenological theory, one can predict the existence of two types of superconducting materials, named type I and type II. Abrikosov and Ginzburg were awarded the Nobel Prize in 2003 for this work (Landau having died in 1968).

The microscopic theory known as the "BCS theory" [12–14] was proposed by John Bardeen, Leon N. Cooper, and John Schrieffer in 1957 (all three received the Nobel Prize in Physics in 1972), partially explaining the fundamental principle of superconductivity. This theory posits that at very low temperatures, electrons move in pairs called "Cooper pairs", due to phonons. After 30 years of theoretical and practical research, the critical temperature limit did not exceed 23.2 K for alloys containing Niobium (Nb_3Ge).

The race for higher critical temperatures was reignited on January 27, 1986, when Johannes Bednorz and Karl Muller discovered superconductivity at a temperature of 35 K in perovskite-structure materials based on lanthanum, barium, copper, and oxygen (LaBaCuO), a discovery for which they received the Nobel Prize in Physics in 1987. Nine months later, Wu et al achieved a critical temperature of 92 K by replacing lanthanum with Yttrium ($YBa_2Cu_3O_7$), surpassing the temperature of liquid nitrogen (77 K, $-196\,°C$), which is 10 times cheaper than liquid helium and cools 20 times better. This sparked a race for high-temperature superconductors, leading to the creation of superconductors exceeding 100 K in 1988.

In 1995, a reproducible critical temperature record was set with mercury-based compounds at a temperature of 164 K under high pressures. The hope for room-temperature superconductors remains alive.

Figure 1.2 shows a histogram of the evolution of critical temperatures of super-conductors over time. We distinguish two types of superconductors: low critical temperature superconductors $T_c < 25$ K and high critical temperature superconductors $T_c > 25$ K (Table 1.1).

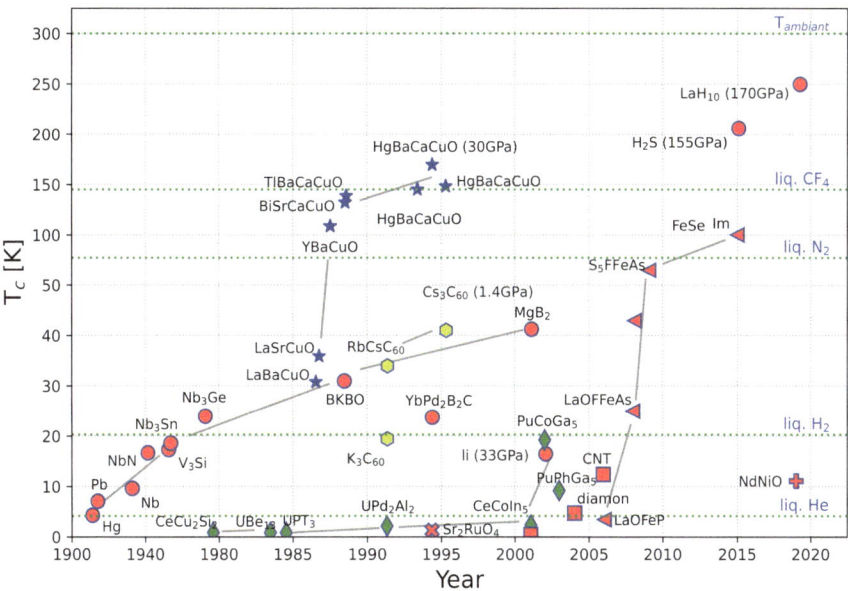

Fig. 1.2 Histogram of significant discoveries in the field of superconductivity (based on [15])

Table 1.1 Superconducting transition temperatures and critical magnetic fields of various compounds

Compounds	T_c (K)	B_c (T)	Ref.
Al	1	0.01	[16–18]
Pb	7.2	0.08	[17, 18]
Nb	9.2	0.2	[17, 18]
Nb_3Sn	18	24	[19]
Nb_3Ge	23	38	[20]
$PbMo_6S_8$	15	60	[21]
$La_{2-x}Sr_xCuO_4$ (1986)	38	40	[22]
$YBa_2Cu_3O_{6.9}$	92	>100	[23]
$Bi_2Ca_2Sr_2Cu_3O_{10}$	110	>120	[24]
$Tl_2Ca_2Ba_2Cu_3O_{10}$	125	>130	[25]
MgB_2 (2001)	39	>50	[26]

1.1.2 History

The phenomenon of superconductivity, discovered and studied over the centuries, has revolutionized our understanding of quantum physics and opened doors to innovative technological applications. The timeline of discoveries in this field reflects an unceasing quest to comprehend and harness this mysterious state of matter.

In 1908, Kamerlingh Onnes achieved a groundbreaking feat by liquefying helium at 4.2 K ($-268.95\,°C$), setting the stage for experiments at very low temperatures [27]. This achievement was pivotal because it provided the essential conditions for studying matter in states that were previously inaccessible.

Three years later, in 1911, Onnes made another monumental discovery: superconductivity. By proposing to G. Holst to measure the resistivity of mercury at very low temperatures, they observed that its electrical resistance abruptly dropped to zero below a critical temperature [28, 29]. This discovery was not only unexpected but also opened an entirely new field of research in condensed matter physics.

The journey into the depths of superconductivity was fraught with challenges and setbacks. In 1913, an attempt to create the first superconducting magnet ended in failure. However, these early experiments laid the groundwork for understanding the delicate conditions under which superconductivity could exist. By 1914, Onnes observed that the superconducting state could be destroyed by applying a magnetic field, introducing the concept of critical magnetic fields in superconductors.

The interaction between superconductivity and magnetic fields became clearer with the work of Sillsbee [30], who demonstrated that applying an electric current could also disrupt superconductivity. This finding hinted at the complex relationship between superconductivity, electric currents, and magnetic fields.

A significant advancement came in 1933 when Meissner and Ochsenfeld discovered the diamagnetism of superconductors [8], known as the Meissner effect. This effect, where a superconductor expels a magnetic field from its interior, was crucial in understanding the true nature of superconductivity. It suggested that superconductivity was not merely about the absence of electrical resistance but involved a complete expulsion of magnetic fields, indicating a new state of matter.

The concept of penetration depth, proposed by Fritz and Heinz London between 1934 and 1935, further elucidated the Meissner effect [31, 32]. They introduced the idea that a magnetic field could penetrate a superconductor but only to a limited depth, providing a theoretical framework to describe the electromagnetic properties of superconductors.

Throughout the mid-twentieth century, the field of superconductivity expanded rapidly. The discovery of type II superconductors by Abrikosov [33]; the formulation of the BCS theory by Bardeen, Cooper, and Schrieffer [34, 35]; and the prediction and observation of the Josephson effects in the 1960s were monumental achievements [36–38]. These advances not only deepened our understanding of superconductivity but also led to practical applications, such as MRI machines, particle accelerators like the Large Hadron Collider, and the development of superconducting magnets.

The discovery of high-temperature superconductors, starting with the work of Bednorz and Müller in 1986, marked a new era in superconductivity research [39, 40]. This discovery challenged the previously held belief that superconductivity could only occur at temperatures close to absolute zero. The subsequent discoveries of superconductors with critical temperatures above the boiling point of liquid nitrogen opened the possibility of more accessible and economically viable applications.

The quest for room-temperature superconductivity remains one of the holy grails of physics. Recent experiments, such as those involving the compression of hydrogen sulfide to achieve superconductivity at temperatures as high as 190 K under high pressure, hint at the potential of discovering materials that could operate as superconductors at even higher temperatures [41].

The timeline of superconductivity is a testament to the relentless pursuit of knowledge and the application of scientific principles to unravel the mysteries of the quantum world. Each discovery builds on the work of those who came before, pushing the boundaries of what is possible and paving the way for future innovations that could transform technology and society.

1.1.3 Applications

The technological interest in superconductors is undeniable. The advantages of high critical temperature superconducting materials include

- The absence of resistivity in superconductors, which prevents any energy dissipation, allows for increased power without raising the consumption of the installation.
- Due to the elimination of the Joule effect (heating due to the material's resistivity), it becomes possible to pass very high current densities through very compact windings.
- Superconductivity enables the generation of very strong magnetic fields.

There are numerous applications across various fields (environment, energy, electricity storage, medicine, transportation, etc.).

Environmental Field: Water purification using superconductivity by exploiting the magnetic properties of superconductors (used in wastewater treatment plants (WWTP) to filter water). Indeed, magnetic particles that attach to dust can be mixed with water to capture dust, then separated from the liquid medium by a powerful magnetic field generated by superconductors, carrying the impurities with them. Superconductors can also be used for air purification, though the process differs from water purification since the goal is different. Here, the aim is to eliminate pollutants before they are released into the environment. For example, during coal combustion, pollutants contained in it (mainly sulfur) are expelled into the air with other smoke species. However, these compounds, lacking the same magnetic properties, can be separated before combustion thanks to a superconducting magnet.

Energy Field: Current transport from upstream to downstream is done through copper or aluminum cables. These metals have disadvantages, namely, the presence of resistance that leads to significant energy loss in the form of heat (Joule effect). An effective solution to avoid Joule effect losses is the use of superconducting cables, which allow for much greater current flow than a conventional line due to their zero resistance. These cables, capable of transmitting very high currents (up to 5 kA), provide an original solution for solving electric power transport problems by increasing intensity rather than voltage. Thus, 8400 Kg of copper cable could be replaced by

110 Kg of superconducting cable. However, the cooling system for these cables remains an obstacle to their market deployment.

Electricity Storage Field: The energy sector has adopted modern electric energy storage systems based on superconductivity. The principle of magnetic energy storage involves passing an electric current through a short-circuited superconducting material coil cooled below its critical temperature [13, 42, 43]. The current will be stored indefinitely as there is no loss since it circulates in the superconductor (kept below the critical temperature). The energy stored in the coil is given by $\frac{1}{2}LI^2$, where L is its inductance and I is the current passing through it.

Medical Field: Superconductivity plays a crucial role in medical imaging. The technology of MRI (magnetic resonance imaging) relies entirely on the power of the electromagnet, which is the very basis of the device. Hundreds of kilowatts would be required for a conventional magnet to reach the desired magnetic field. Consequently, the use of a superconducting magnet is highly advantageous. MRI is based on the principle that hydrogen atoms in the water molecules, which make up the majority of the human body (about 70% of it, targeted in MRI), excite when subjected to magnetic fields due to superconducting magnets and thus produce energy which they return by emitting a signal that can be detected. Thanks to the development of computing and signal processing, the vibrations sent back allow for the creation of an image of the inside of the body. Superconductors appear at two points: firstly, in the electromagnet that induces the strong magnetic field necessary for the atoms to accumulate energy, and, secondly, in the electromagnetic wave detector, which provides the image of the brain [44]. Very intense (0.5–4 Tesla) and very stable magnetic fields are necessary for medical imaging to improve image resolution. They can be achieved through superconducting magnets.

1.2 Theoretical Overview

1.2.1 Superconducting Materials

Superconductivity transcends conductivity as it entirely disregards the notion of conductivity. Among the new properties that emerge below T_c (the critical temperature), we can distinguish those that involve the very nature of the new ground state, including the absence of resistivity and the Meissner effect.

1.2.2 Electrical Resistivity

A superconductor is a material that has zero electrical resistivity below a certain critical temperature. T_c is the transition temperature that delineates the superconducting state from the normal state (Fig. 1.3).

Fig. 1.3 Typical curve of the resistivity of a superconductor as a function of temperature

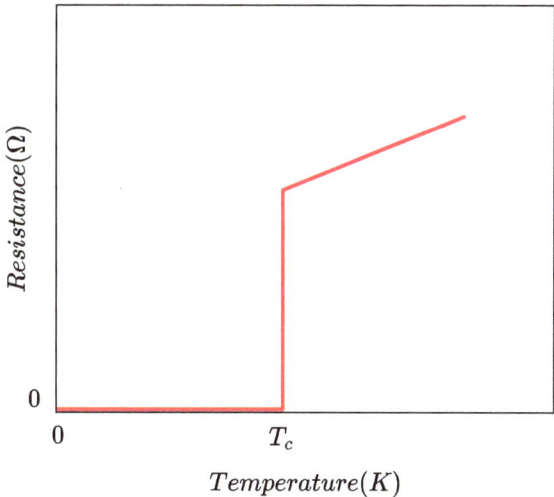

1.2.3 Meissner Effect or Perfect Diamagnetism

The disappearance of resistivity is accompanied by an exceptional magnetic property known as the Meissner effect, which entails the complete exclusion of the magnetic field within the material when it is in the superconducting state. In 1933, Meissner and Ochsenfeld [47] discovered that when a superconductor is cooled below its critical temperature and then placed in a magnetic field, the magnetic flux lines cannot penetrate the interior of the sample (Fig. 1.4). An external magnetic field induces the circulation of currents on its surface, which in turn create a magnetic field that

Fig. 1.4 Effect of the magnetic field on a superconducting material in the Meissner state $(T < T_c)$ and in the normal state $(T > T_c)$

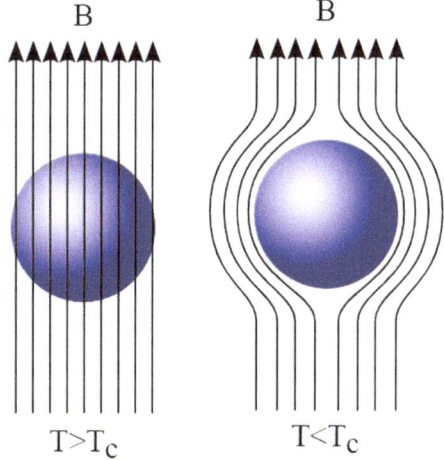

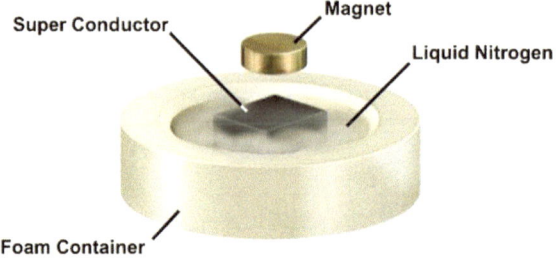

Fig. 1.5 Meissner effect

opposes the passage of the magnetic field to which it is subjected. The susceptibility $\chi = -1$ in the sample.

To clarify this, one can place a magnet above a superconductor and observe that the magnet levitates above the superconductor (Fig. 1.5).

1.2.4 Types of Superconductors

1.2.4.1 Type I Superconductors

Type I low-temperature superconductors primarily consist of pure metals (Pb, Hg, In, Sn, Al,...), or alloys such as (NbN, V_3Si, Nb_3Ge,...). They are characterized by a single critical field threshold, $H_c(T)$. Superconductivity completely disappears when a magnetic field greater than H_c is applied (Fig. 1.6). Indeed, for $T < T_c$:

$$\vec{B} = \mu_0(\vec{H} + \vec{M}) = 0 \quad \Rightarrow \quad \vec{M} = -\vec{H} \tag{1.1}$$

where $\vec{M}$ is the magnetization, $\vec{H}$ is the magnetic field inside the material, and μ_0 is the vacuum permeability. Thus, these superconductors are perfect diamagnetics

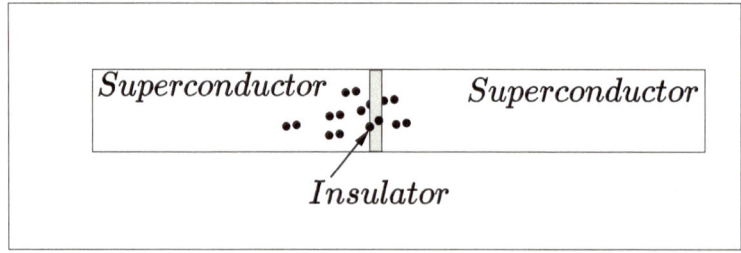

Fig. 1.6 a Variation of the critical field H_c as a function of temperature. **b** Magnetization curve M as a function of the applied magnetic field H at a given temperature (for type I superconductor [45, 46])

($\chi = -1$) when $H < H_c$. The sample returns to its normal phase when $H > H_c$, the Meissner effect disappears, and the applied magnetic field fully penetrates the sample.

The value of the critical magnetic field H_c varies as a function of the sample's temperature T and the critical temperature T_c of the material, following a function of the type:

$$H_c(T) = H(0)\left[1 - \left(\frac{T}{T_c}\right)^2\right] \tag{1.2}$$

where $H(0)$ is the value of the magnetic field at $0\,$K. (This yields the phase diagram shown in Fig. 1.6.) The value of the critical field H_c is always very low for type I superconductors, which limits their practical use.

1.2.4.2 Type II Superconductors

Type II superconducting materials possess two critical fields, H_{c1} and H_{c2}. The value of the magnetic field H_{c2} is greater than that of H_{c1} [48, 49].

The behavior of a type II superconductor is identical to that of a type I superconductor as long as $H < H_{c1}$, it is perfectly diamagnetic. Beyond this, diamagnetism becomes partial (the Meissner effect is no longer total), and the material is then in the mixed or vortex state (Fig. 1.7). Above the value H_{c2}, the material is entirely in the normal state.

The phase diagram of type II superconductors (Fig. 1.7) primarily includes alloys or metals that have a high resistivity in the normal state and elevated critical fields H_{c2}.

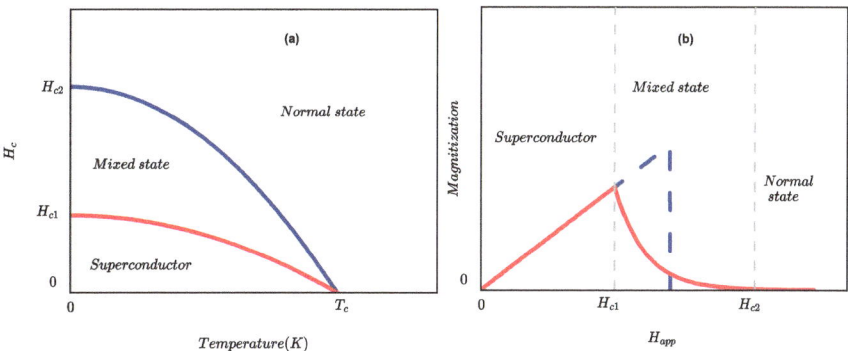

Fig. 1.7 a Variation of the critical fields H_{c1} and H_{c2} as a function of temperature. **b** Variation of the magnetization M in function of the magnetic field H (for type II superconductor [45, 46])

1.3 Phenomenological Theories

1.3.1 Penetration Depth: London Theory

The theory of the London brothers, Fritz and Heinz, represents the first phenomeno-logical approach to superconductivity and was proposed in 1935 [10, 50]. This theory describes the Meissner effect and is based on two phenomenological equations that govern local fields. The two London equations, which relate the microscopic electric fields $\vec{E}$ and microscopic magnetic fields $\vec{H}$ with the superconducting electron current density $\vec{J}_s$, are written as follows:

$$\vec{E} = \mu_0 \lambda_L^2 \frac{\partial \vec{J}_s}{\partial t} = \frac{m_e}{n_s e^2} \frac{\partial \vec{J}_s}{\partial t} \tag{1.3}$$

And the magnetic field $\vec{H}$ is related to the superconducting current density $\vec{J}_s$ through the equation:

$$\vec{H} = -\lambda_L^2 \nabla \times \vec{J}_s \tag{1.4}$$

where

- n_s is the density of superconducting electrons.
- J_s is the superconducting current density.
- m_e is the electron mass.
- λ_L is the London penetration depth, indicating how deeply magnetic fields can penetrate into a superconductor.

Using the Maxwell–Ampère equation, and with neglecting displacement current, Eq. (1.4) leads to Eq. (1.5).

$$\nabla^2 \vec{H} - \frac{1}{\lambda_L^2} \vec{H} = 0 \tag{1.5}$$

The solution of this differential equation shows that the screening current and the magnetic field in a superconductor subjected to an applied magnetic field decay exponentially (B and $J \propto \exp(-r/\lambda_L)$) from the external surface toward the interior. This implies that the external magnetic field is screened inside the sample over a characteristic thin thickness λ_L (Fig. 1.8), called the London penetration depth.

$$\lambda_L(T) = \sqrt{\frac{m_s}{\mu_0 q_s^2 n_s(T)}} = \lambda_L(0) \sqrt{\frac{1}{1 - t^4}} \tag{1.6}$$

where q_s and m_s are, respectively, the charge and mass of the superconducting elec-trons, $n_s(T)$ becomes zero at $T = T_c$ and maximal at $T = 0$ ($n_s = n$, with n being the sum of the volumetric densities of normal (n_n) and superconducting (n_s) elec-trons), and the ratio $t = T/T_c$ corresponds to the reduced temperature. λ_L increases

Fig. 1.8 Density of superconducting electrons and penetration of the magnetic field into a superconductor over a length λ_L

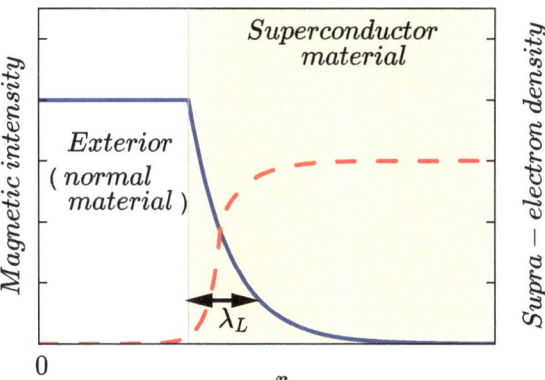

as the temperature rises and tends toward infinity at $T = T_c$, naturally leading to the penetration of the magnetic field beyond the critical temperature.

1.3.2 Ginzburg–Landau Theory

The Ginzburg–Landau theory was proposed in 1950 by Soviet physicists Ginzburg and Landau [51, 52]. It is based on Landau's postulate, which expresses the free energy of the superconducting system. This expression, introduced to describe second-order thermodynamic phase transitions, posits that the superconducting state is characterized by a complex order parameter $\psi(r)$, representing a sort of "effective wave function of the superconducting electrons" that depends on space. The physical interpretation is that its squared modulus represents the density of charge carriers in the superconducting state (Cooper pairs) ($|\psi(r)|^2 = n_s(r)$). Near T_c and in the absence of a magnetic field, the order parameter $|\psi(r)|^2 = $ const and the volumetric free energy difference between the superconducting and normal phase leading to an equilibrium state can be reduced to these first two terms:

$$f_s(T) - f_n(T) = \alpha(T)|\psi(r)|^2 + \frac{1}{2}\beta(T)|\psi(r)|^4$$
$$= -\frac{1}{2}\mu_0 H_c^2 \tag{1.7}$$

where

- $f_n(T)$ is the free energy density in the normal state.
- $f_s(T)$ is the free energy density in the superconducting state.
- α and β are coefficients of the expansion of f_s in powers of ψ, characteristic parameters of the sample that depend only on temperature.

The simplest analytical form of $\alpha(T)$ leading to a zero-order parameter above T_c and non-zero below this temperature is $\alpha(T) = \alpha(T - T_c)$, and the $\beta(T)$ parameter must be positive ($\beta = \text{constant} > 0$).

- If $\alpha(T) > 0$, $|\psi| = 0$ thus $f = f_n \Rightarrow$ the system is then in the normal state.
- If $\alpha(T) < 0$, $|\psi| = \pm\sqrt{\alpha(T)/\beta} = \psi_0$ and $\mu_0 H_c^2 = \alpha^2/\beta \Rightarrow$ the system is then in the superconducting state.

The introduction of a volumetric kinetic energy term in Eq. (1.7) allows for the spatial variation of the order parameter to be taken into account and obtains the Ginzburg–Landau penetration length:

$$\lambda_{GL}(T) = \sqrt{\frac{m_s}{\mu_0 q_s^2 |\psi_0|^2}} = \sqrt{\frac{m_s}{\mu_0 q_s^2 n_s(T)}} \tag{1.8}$$

The length λ_{GL} coincides with the London penetration length λ_L and corresponds to the layer over which the applied magnetic field penetrates the superconducting sample, resulting in a gain of magnetic energy.

Additionally, the Ginzburg–Landau equation defines the coherence length $\xi_{GL}(T)$ as the scale of spatial variations of the order parameter within the superconductor. It represents the shortest distance over which ψ can vanish, and is expressed as

$$\xi_{GL}(T) = \sqrt{\frac{\hbar^2}{2m_s |\alpha|}} \tag{1.9}$$

with

$$\alpha = -\left(\frac{\mu_0^2 q_s^2}{m_s}\right) H_c^2(T)\lambda_{GL}^2(T) \tag{1.10}$$

The parameters m_s, n_s, and q_s correspond, respectively, to the mass, density of elementary carriers, and the charge ($q_s = -2e$).

As both λ_{GL} and ξ_{GL} are inversely proportional to α, they depend on temperature and both diverge as T approaches T_c. Their ratio $\kappa = \lambda_{GL}(T)/\xi_{GL}(T)$, called the Ginzburg–Landau parameter, allows distinguishing two cases (Fig. 1.8):

- If $\kappa < 1/\sqrt{2}$, the interface's surface energy between the normal and superconducting phases is positive. Thus, the behavior observed by Meissner is seen, i.e., the flux is entirely excluded from the material below T_c, superconductivity is destroyed, and the material returns to the normal state. In this case, the material is known as type I.
- If $\kappa > 1/\sqrt{2}$, however, the surface energy of a normal/superconductor interface is negative, and it will therefore be energetically favorable for flux to exist within the superconducting material, and, in this case, the superconducting material is said to exhibit type II behavior.

1.3.3 Abrikosov's Theory of the Mixed State in Type II Superconductors

1.3.3.1 Introduction

Based on the Ginzburg–Landau theory, Abrikosov [54, 55] considered in 1957 the case of negative wall energies in the mixed state of a type II superconductor. In this state, the distribution of induction is not uniform. Indeed, magnetic induction appears only at various points. These points form a triangular lattice (Fig. 1.9).

The points are, in fact, magnetic flux tubes that penetrate the superconductor as elementary entities parallel to the magnetic field. These structures, called vortices, each carry a quantum multiple of flux ϕ_0, with this elementary flux given by the expression:

$$\phi_0 = \frac{h}{2e} = 2.07 \times 10^{-15}\,\text{Wb} \tag{1.11}$$

where h is Planck's constant and e is the charge of the electron. Thus, the center of the vortex, also called the vortex core, is always in the normal state, while the rest of the material is in the superconducting state. As the external field increases, the number of lines (n) per square centimeter increases, and the magnetic induction is $B = n\phi_0$.

For $H_{c2} > H > H_{c1}$ (Fig. 1.10), the magnetic field begins to partially penetrate the superconductor in the form of isolated flux lines or vortices because superconducting currents circulate around the axis of each of them, screening the magnetic field beyond a distance on the order of λ. B is maximal at the center of the tube. The order parameter has its amplitude $|\psi_\infty|$ except near the (almost normal) core of the vortex, with a radius of approximately ξ, where it decreases to zero along the axis.

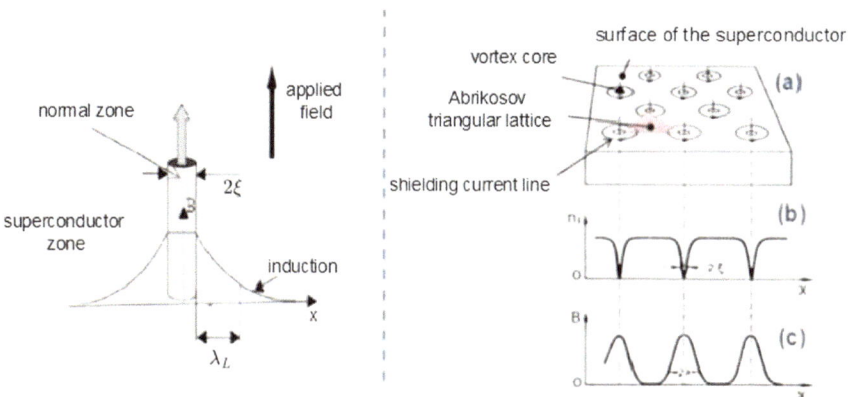

Fig. 1.9 On the right, formation of a vortex. On the left, **a** Mixed state in an applied field just above H_{c1}, **b** Density of superconducting electrons, and **c** Variation of the flux density [53]

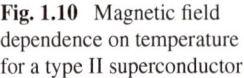

Fig. 1.10 Magnetic field dependence on temperature for a type II superconductor

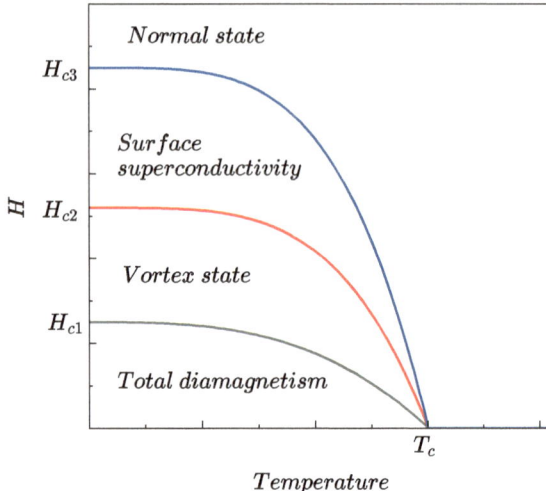

A vortex is surrounded by a supercurrent of density j_s that screens the magnetic field outside. This current flows within a thickness of value λ and has a radius equal to the coherence length ξ. The vortices repel each other and thus arrange themselves in a regular organization leading to a minimum energy level, often in a triangular lattice.

For $H_{c3} > H > H_{c2}$, a superconducting layer forms, in which only a thin surface portion of the material is superconducting (and this only if the magnetic field has a component parallel to the material's surface).

1.3.3.2 Surface Energy

The existence of finite lengths $\lambda(T)$ and $\xi(T)$ has numerous experimental consequences for describing the behavior of superconductors in an applied magnetic field, which can often be expressed using the notion of surface energy σ_{NS} between the superconducting state and the normal state. This energy has an energetic cost on the order of $(\mu_0 H_c^2/2)(\xi - \lambda)$. σ_{NS} plays a fundamental role in the physics of superconductors. It is particularly shown that this surface energy is not always positive, and the sign of σ_{NS}, which is related to the Ginzburg–Landau constant κ, allows for classifying superconductors into two major classes:

- $\sigma_{NS} > 0$ if $\kappa < 1/\sqrt{2}$: we have a type I superconductor.
- $\sigma_{NS} < 0$ if $\kappa > 1/\sqrt{2}$: we have a type II superconductor.

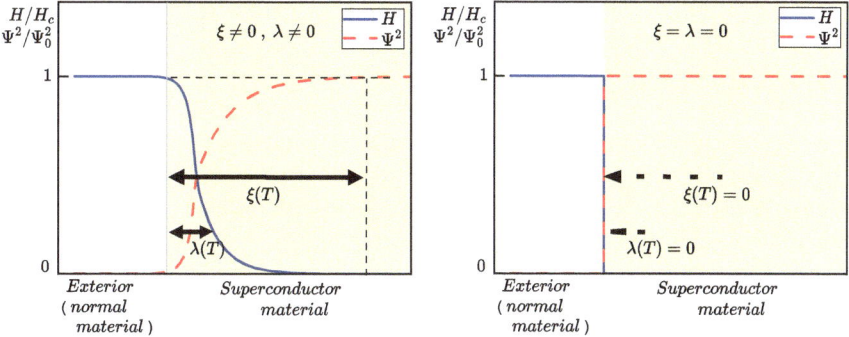

Fig. 1.11 Variation of the order parameter $|\psi|$ (dashed line) and the magnetic field H (solid line) near an S-N (Superconductor-Normal) interface

Figure 1.11 illustrates the variation of the order parameter $|\psi|$ and the magnetic field H near an N-S (normal–superconductor) interface.

- ψ^2 and H change abruptly at the N/S interface ($\lambda = \xi = 0$).
- ψ^2 and H vary over their characteristic distances $\lambda(T)$ and $\xi(T)$.

We note that the sign of the surface energy has no influence on the transition temperature.

1.3.3.3 Properties of the Isolated Vortex

Solving the modified London equation gives the local magnetic field $H(r)$ and the associated current $J(r)$. In the case of an isolated vortex, these parameters decrease exponentially from the center of the vortex over the distance λ. $n_s = |\psi|^2$ is zero at the core of the vortex (Fig. 1.9b) and returns to the value $|\psi_\infty|^2$ it had at $B_c < B_{c1}$ in the absence of the vortex (Fig. 1.9c). Thus, a vortex is a whirl of superconducting current around a normal metallic tube with a radius $\xi \ll \lambda$. This current screens the vortex core from the surrounding superconducting regions and allows each vortex to carry a flux equal to the quantum of flux or fluxoid ϕ_0. There is no chemical or crystallographic difference between the core and the rest of the vortex. The vortex state ($H_{c1} < H < H_{c2}$) is stable because the penetration of the applied magnetic field into the superconducting material renders the surface energy negative [56, 57]. For $H \geq H_{c1}$, the field at the core of the vortex is proportional to H_{c1}. It decreases to zero over the length λ. Its flux $\pi\lambda^2 H_{c1}$ is approximately equal to ϕ_0, hence $H_{c1} \approx \phi_0/\pi\lambda^2$. For $H \leq H_{c2}$, the vortices are tightly packed together, and each vortex occupies an area approximately $\pi\xi^2$, hence $H_{c2} \approx \phi_0/\pi\xi^2$.

1.3.4 Conventional Theory of Superconductivity (BCS)

1.3.4.1 Introduction

The Ginzburg–Landau theory does not explain the microscopic physical origin of superconductivity, but it does allow for the phenomenological description of the macroscopic manifestations of the superconductivity phenomenon. In 1957, American physicists J. Bardeen, L. Cooper, and J. Schrieffer proposed the BCS (Bardeen–Cooper–Schrieffer) theory to explain the microscopic mechanisms behind superconductivity [14], for which they received the Nobel Prize in Physics in 1972. This theory is based on the idea that two electrons can pair to form a Cooper pair, thus behaving as a boson rather than two fermions. They form a unique, coherent state of lower energy than that of the normal metal (unpaired electrons). The formation of a Cooper pair requires an electron–electron attraction stronger than the Coulomb repulsion. This attraction is mediated by phonons.

A Cooper pair can be destroyed by

- Electron–electron Coulomb repulsion: Not all materials are superconductors.
- Thermal agitation: The superconducting state exists only below the critical temperature.
- The action of a magnetic field: Superconductivity is lost beyond a critical field. The loss can be abrupt (type I superconductor) or gradual (type II superconductor, in which magnetic field vortices can penetrate).

The energy difference between the superconducting and normal states is called the energy gap and is denoted by Δ. It is the energy required to transition from the superconducting to the normal state by breaking Cooper pairs. It tends to zero as the temperature T approaches the critical temperature T_c (Fig. 1.12).

Fig. 1.12 Temperature dependence of the amplitude of the superconducting order parameter. It is zero at the critical temperature T_c and then gradually increases as the temperature decreases

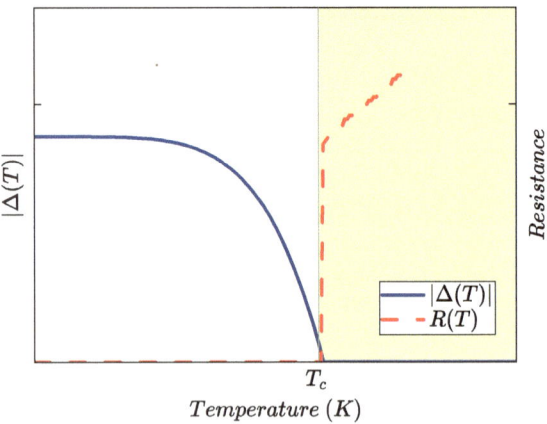

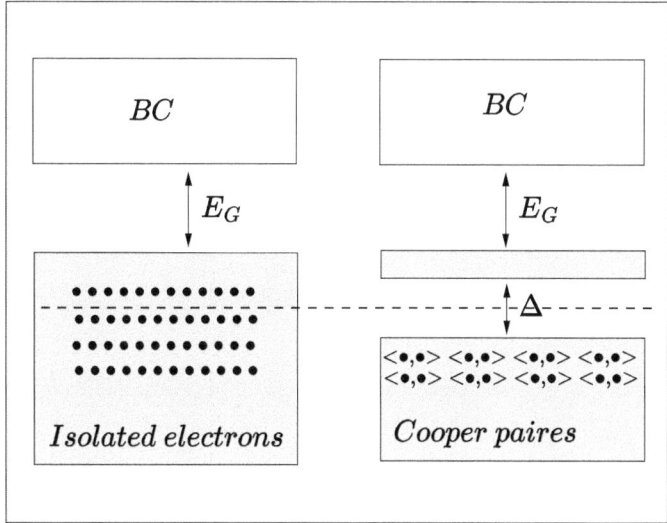

Fig. 1.13 Electronic filling for a degenerate semiconductor. Right: superconducting case. A forbidden band of width Δ appears. Left: non-superconducting case. The valence band (VB) is filled up to the Fermi level (dashed line). The conduction band (CB) is empty

The existence of this superconducting energy gap of width Δ (Fig. 1.13) naturally appears in this theory, which specifies its evolution as a function of temperature, particularly the relation: $\Delta(0) = 1.76 \, k_B T_c$. To break a Cooper pair, one must then provide an energy of 2Δ.

1.3.4.2 Origin of Attractive Interactions—Cooper Pairs

When electrons pass through a crystalline lattice formed of positive ions, they locally attract the ions and slowly converge among themselves, forming atomic vibrations called phonons. The interaction between electrons and phonons is the origin of resistivity and superconductivity. The movement of ions forms a zone of positive charge. If this is greater than the charge of an electron, it can attract another passing electron (Fig. 1.14). Hence, the formation of an electron pair is called a Cooper pair, which leads to the existence of a coherence length ξ.

At high temperatures, the ions of the material are in motion, agitating, and the equilibrium is then broken, and the Cooper pairs disappear. The BCS theory explains well that at low temperatures, electrons undergo no energy dissipation by the Joule effect because they can pass through the crystalline lattice without encountering any obstacle. Since the first electron clears a path for the second, which is drawn in without friction against the particles of the material, which is obviously not the case when these are agitated. Therefore, the BCS theory applies only to materials that are superconductors at low temperatures on the order of 30 K.

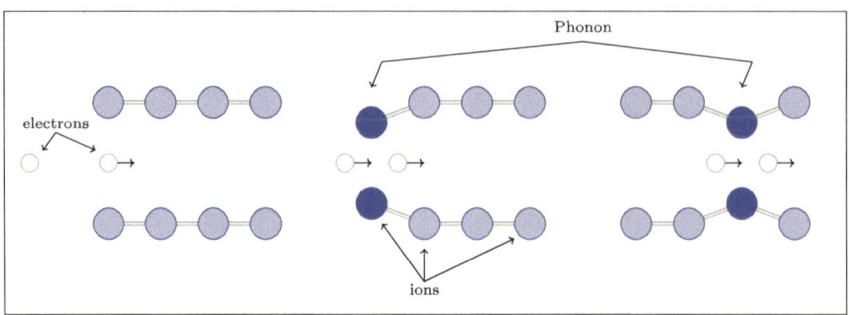

Fig. 1.14 Superconducting state of a material

1.3.4.3 Results of BCS Theory

Energy Gap The energy gap separates the superconducting electrons below from the normal electrons above (Fig. 1.13). In the weak coupling limit between electrons, the superconducting gap is related to the critical temperature by the relation:

$$k_B T_c = 1.14 \hbar \omega_D \exp\left(-\frac{1}{N(E_F)V}\right) \tag{1.12}$$

$$\frac{E_g}{2} = \Delta(0) \approx 2\hbar \omega_D \exp\left(-\frac{1}{N(E_F)V}\right) \tag{1.13}$$

where k_B is the Boltzmann constant, ω_D is the Debye frequency, $N(E_F)$ is the electronic density of states at the Fermi level, and V is the interaction potential of the two electrons in a Cooper pair. From this, we can derive the famous equation:

$$EG = 2\Delta = 3.53 \, k_B \, T_c \tag{1.14}$$

This gap is confirmed for classical superconductors (for example, $2\Delta / k_B T_c = 3.5$ for Sn, 3.8 for Nb) [58]. The critical field, thermal properties, and most electromagnetic properties directly stem from the existence of this gap. In some cases, superconductivity can appear without an energy gap. The length of this gap decreases as the temperature increases, until it vanishes at the critical temperature.

$$\Delta(T) \sim T_c \left(1 - \frac{T}{T_c}\right)^{1/2} \tag{1.15}$$

Equation (1.12) shows that for $T_D = (\hbar \omega_D)/k_B \approx 100$ to 500 K and $V N(E_F) \approx 0.3$, we have $T_c \approx 25$ K. Thus, the BCS theory only explains the T_c of conventional superconductors.

Coherence Length For a clean metal, the characteristic length over which the pairing formation evolves spatially is given by $\xi_0 = \frac{\hbar v_F}{\pi \Delta}$. Here, ξ_0 is then the coherence length of the superconducting metal, representing the distance between the two electrons in a Cooper pair. The typical order of magnitude of ξ_0 is from hundreds of nanometers to one micrometer in metals like aluminum or niobium, which is large compared to the distance between electrons.

1.4 Josephson Junctions

A Josephson junction consists of an extremely thin insulating layer, on the order of nanometers, separating two superconductors (SIS) (Fig. 1.15). The Josephson effect was discovered by Josephson [59, 60]. This effect results from the tunneling of "Cooper pairs" rather than individual electrons. By spontaneously passing from one superconductor to the other, the pairs create a continuous electric current. This current can propagate without resistance between the two superconducting zones. These currents are called "Josephson currents", and the non-conductive material, called a "Josephson Junction", then behaves like a superconductor.

Strictly speaking, the Josephson effect has its limits. Indeed, the direct current flowing between the two superconductors must be very low for the tunneling effect to be observed. If the current being passed is greater than the critical current, then the Josephson junction behaves as an obstacle, and the resistance, which has been null up to now, becomes very significant.

When an alternating current flows between the two superconductors, the current in the Josephson junction oscillates at a frequency that depends only on the applied voltage u, aside from a constant. The expression for the frequency of this oscillation is given by the relation: $f = \frac{2e}{h} \cdot u$. With e the charge of the electron and h Planck's constant.

The reciprocal effect can also be obtained: if a Josephson junction is subjected to a variable magnetic field, then an alternating current is observed on either side of the two superconductors, verifying the same relationship as above.

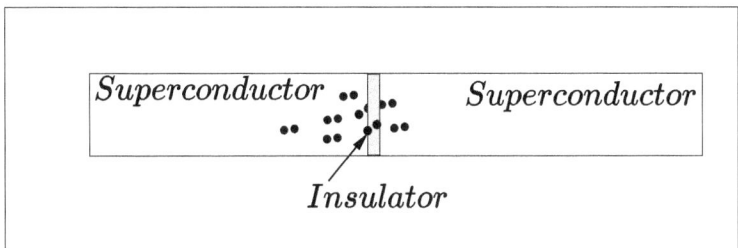

Fig. 1.15 Josephson junction

1.5 High Critical Temperature Superconductors

1.5.1 History and Progress

The BCS theory predicted that the phenomenon of superconductivity could not exist beyond 25–30 K, which had the effect of slowing the quest for new superconductors. However, everything changed in 1986 with the discovery of the first high critical temperature superconductor in a copper perovskite-type ceramic $La_{2-x}Ba_xCuO_{4-\delta}$ (with a critical temperature $T_c = 35$ K) by the German chemist J. Georg Bednorz and the Swiss physicist Alex Müller [61], at the IBM research laboratory in Zurich, for which they won the Nobel Prize in Physics the following year. Since then, a race for high critical temperature superconductors began, and discoveries followed rapidly:

- In 1987, the temperature barrier of liquid nitrogen was passed, and superconductivity appeared at 92 K for $YBa_2Cu_3O_{7-\delta}$ [62]. Compounds of the same family were also studied by substituting yttrium with rare earths (neodymium, dysprosium).
- Then, in 1988, 110 K was achieved for the compound $Bi_2Sr_2Ca_2Cu_3O_{10-\delta}$ [45]. That same year, a T_c of 125 K was reached for the compound $Tl_2Ba_2Ca_2Cu_3O_{10}$ [63].
- The year 2009 was marked by the discovery of superconductivity in a system (at atmospheric pressure) based on barium, calcium, mercury, and copper oxide $(HgBa_2Ca_2Cu_3O_x)$, at 135 K [64], this temperature eventually reached 164 K under high pressure [65].
- Iron-based superconductors with critical temperatures above 56 K have been discovered [66, 67] (Fig. 1.16).

These compounds, all of the same perovskite type, were discovered subsequently, and each time, the critical superconducting transition temperature was increased.

After about 40 years of intensive research (the goal is the quest for room-temperature superconductivity [69]), the origin of high-temperature superconduc-

Fig. 1.16 Molecular perovskite structure [68]

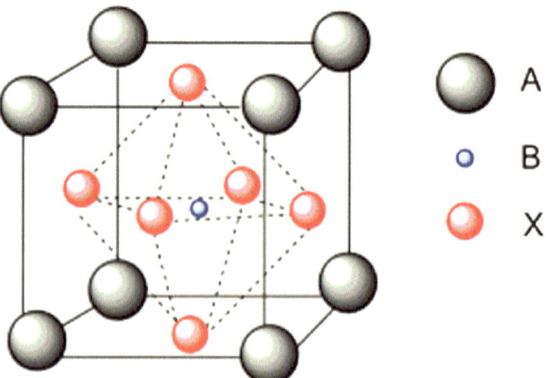

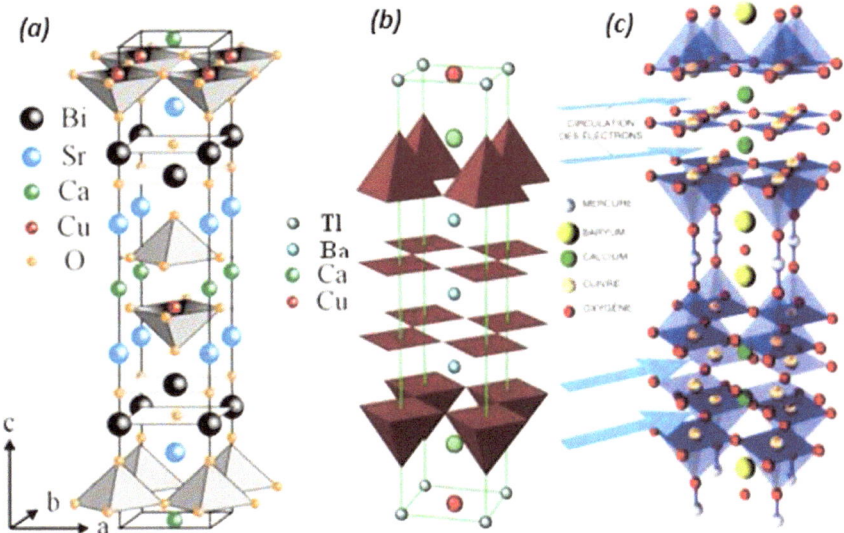

Fig. 1.17 Structure of: **a** $Bi_2Sr_2CaCu_2O_{8+\delta}$, **b** $Tl_2Ba_2Ca_2Cu_3O_{10+\delta}$, and **c** $HgBa_2Ca_3Cu_4O_{8+\delta}$ [70]

tivity is not yet unveiled, but it seems that, unlike the electron–phonon attraction mechanisms that describe well conventional superconductivity, we are dealing with mechanisms of electronic origin.

High critical temperature superconductors differ from "classical" superconductors in that they are oxides, instead of being intermetallic compounds. All high critical temperature superconductors have an elementary cell structure derived from the perovskite structure, formed by layers of CuO_4 tetrahedra between which are the alkaline earth ions (Ca, Ba, Sr), separated by layers containing lanthanides (Y and rare earths) or heavy metals (Hg, Pb, Bi, Tl). This layered structure gives a very strong anisotropy to the mechanical and electrical properties of these compounds (Fig. 1.17).

The fundamental element common to all high-temperature superconductors (HTSC) is the presence of copper–oxygen CuO_2 planes, which are attributed with superconducting properties. The electrons responsible for superconductivity thus move in a two-dimensional square lattice.

For example, the planes containing copper ions have a conductivity in the normal state that is much higher than that of the intermediate planes. Consequently, the resistivity along the c-axis (perpendicular to the CuO_2 planes) can be more than 1000 times greater than the resistivity along the a- and b-axes (within the CuO_2 planes). In the superconducting state as well, one can distinguish the CuO_2 planes with strong superconductivity (high density of Cooper pairs) from the intermediate planes with weak superconductivity (low density of Cooper pairs).

Furthermore, in the polycrystalline state, superconducting oxides are composed of grains joined together by grain boundaries. These can constitute a barrier to current flow, especially along the c-direction where their thickness is greater than the coherence length [71–73]. HTSCs exhibit a highly anisotropic two-dimensional character.

1.5.2 Examples

$La_{1.85}Ba_{0.15}CuO_4$ and YBCO (yttrium barium copper oxide) are well-known examples of high-temperature superconducting cuprates, which is known as the first material to be superconducting above the boiling point of liquid nitrogen. The table below groups some superconducting materials and their critical temperatures below which superconductivity manifests (Table 1.2).

1.5.2.1 Cuprates

High critical temperature superconductors are almost all cuprates, which are crystals whose elementary cell contains an alternation of copper oxide CuO_2, separated by planes made up of sheets of transition ions or other oxides. The CuO_2 planes, where conduction occurs, are inserted between so-called "reservoir" planes containing cations whose nature varies from one compound to another: Y^{3+}, Ba^{2+}, Ca^{2+}, Sr^{2+} [74]. The elementary cell of cuprates is indeed a stacking of blocks of the cubic perovskite type (Fig. 1.18).

There are three characteristics common to all superconducting cuprates:

Table 1.2 Superconducting transition temperatures for various materials [16–20, 23, 26]

Class	T_c (K)	Materials
Superconducting oxides	133	$HgBa_2Ca_2Cu_3O_x$
	110	$Bi_2Sr_2Ca_2Cu_3O_{10}$ (BSCCO)
	90	$YBa_2Cu_3O_7$ (YBCO)
Iron-based superconductors	55	SmFeAs(O, F)
	41	CeFeAs(O, F)
	26	LaFeAs(O, F)
Boiling point of liquid nitrogen—77 K		
Low-temperature superconductors	18	Nb_3Sn
	10	NbTi
	9.2	Nb
	4.2	Hg

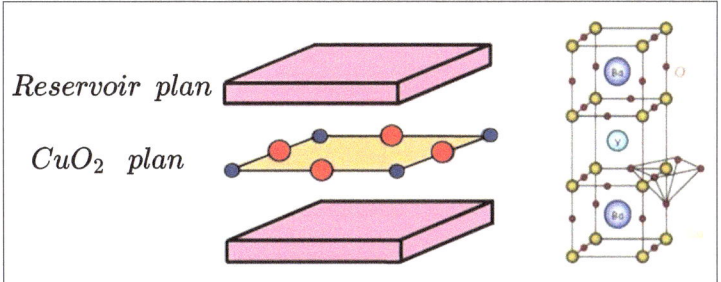

Fig. 1.18 Structure of cuprates (example of YBaCuO)

- Electronic transport occurs preferentially in the CuO_2 planes.
- The CuO_2 planes are alternated with other layers that control the concentration of charge carriers (hole or electron) of these conducting planes. By modifying the chemical composition of these intermediate layers (charge reservoirs), the concentration of charge carriers in the CuO_2 planes is altered.
- The previous operation is called doping, it allows doping the material with holes or electrons to modify their more or less insulating or metallic character [75–77]. Hole-doped cuprates possess the highest known critical temperatures.

1.5.2.2 Iron-Based Superconductors

In 2008, the team of Hideo Hosono from the Tokyo Institute of Technology surprised physicists with the discovery of iron-based superconductors, the pnictides (such as arsenic or phosphate, or a chalcogen element), with critical temperatures nearly as high (maximum T_c is about 55 K) as those in cuprates. The highest critical temperatures in the family of iron-based superconductors exist in thin films of FeSe, where a critical temperature above 100 K was recently reported [78, 78, 79].

Similar to cuprates, the pnictide materials have a layered structure and contain FeAs layers which, to some extent, are analogous to the active CuO_2 planes in cuprates. Within an FeAs layer, each iron atom is placed at the center of a tetrahedron formed by four arsenic atoms. Since the first discoveries, several families of iron-based superconductors have emerged:

- LnFeAs(O, F) or LnFeAsO$_{1-x}$ with a T_c up to 56 K, called 1111 materials [80]. A fluoride variant of these materials was later discovered with similar T_c values [81].
- (Ba, K)Fe$_2$As$_2$ with pairs of iron arsenide layers, called 122 compounds. The T_c values go up to 38 K [82, 83]. Superconductivity is also observed in similar compounds where iron is replaced by cobalt.
- LiFeAs and NaFeAs with a T_c up to about 20 K. These superconducting materials close to stoichiometric composition are considered as 111 compounds [84–87].

1.5.2.3   Materials Sometimes Referred to as High-Temperature Superconductors

Magnesium diboride (MgB_2) is an ionic compound, sometimes considered a high critical temperature superconductor because its T_c is about 39 K, which is higher than historically predicted for BCS superconductors. However, it is more generally considered the conventional superconductor with the highest T_c. The increase in T_c results from two distinct bands present at the Fermi energy [88, 89].

The family of fullerides, where alkali metals are intercalated within C_{60} molecules, as described in [90], demonstrates superconductivity, notably with Cs_3C_{60} reaching temperatures up to 38 K [91–93].

There are organic superconductors and heavy fermion compounds. The first organic superconductor based on fulvalene-type molecules was discovered in 1979 in Orsay. For these materials, consisting of columnar stacking of organic molecules capable of acquiring a positive charge in the presence of an inorganic ion or an electron-accepting molecule, the electrical transport is one-dimensional along the axis of the stacks. One-dimensional organic superconductors offer modest critical temperatures on the order of a degree Kelvin, but a crystallographic variant of these materials in which transport occurs along planes (lamellar systems) has allowed increasing the critical temperature up to 14 K.

Heavy fermions are intermetallic alloys containing rare earth (Cerium) or actinide (Uranium) atoms. It is the electrons "at the bottom" of these atoms that give the exceptional properties to heavy fermions, notably due to their magnetism. Indeed, these electrons through a quantum process entangle (hybridization) with the passing electrons to form a new particle, a sort of electron dressed with the presence of all the others. These new particles, fermions, are significantly slowed down and behave as much heavier particles, as if they were stuck in a massive traffic jam of electrons.

All known high-temperature superconductors are type II superconductors. Unlike type I superconductors, which expel magnetic fields due to the Meissner effect, type II superconductors allow magnetic fields to penetrate in quantized units of flux, creating "holes" or "tubes" of normal metallic regions within the superconductor volume. Therefore, high-temperature superconductors can support much higher magnetic fields.

Magnesium diboride (MgB_2) is often classified as a high T_c superconductor due to its critical temperature of about 39 K, surpassing the limits predicted by the BCS theory for superconductors. Nonetheless, it is widely acknowledged as the conventional superconductor with the highest critical temperature. The enhancement in T_c is attributed to the existence of two distinct bands at the Fermi level.

The family of fullerides, as described in [90], demonstrates superconductivity, notably with Cs_3C_{60} reaching temperatures up to 38 K [91], where alkali metals are intercalated within C_{60} molecules.

Organic superconductors and heavy fermion materials also contribute to the landscape of superconductivity. The inaugural organic superconductor, utilizing fulvalene-type molecules, was unveiled in 1979 in Orsay. These materials feature a 1-D electrical conduction mechanism along the columnar stacks of organic molecules that gain a positive charge by either an inorganic ion or an electron-accepting entity. Although the critical temperatures of one-dimensional organic superconductors are modest, typically around a few Kelvins, a crystallographic variation enabling planar (lamellar) transport has elevated the critical temperature to as high as 14 K.

Heavy fermions, comprising intermetallic alloys with rare earth (Cerium) or actinide (Uranium) elements, owe their remarkable properties to the deep-lying electrons of these atoms. These electrons engage in a quantum hybridization with conduction electrons, resulting in the emergence of heavily massed new fermions, mimicking electrons engulfed in a dense electron traffic jam, which is central to their magnetic characteristics.

It's noted that all high-temperature superconductors fall into the category of type II superconductors, distinguishing themselves from type I superconductors, which completely repel magnetic fields through the Meissner effect. Type II superconductors permit magnetic fields to infiltrate in quantized flux units, forming "holes" or "tubes" of normal conductive areas within the superconducting matrix, enabling them to withstand significantly higher magnetic fields.

1.6   Conclusion

The journey through the fascinating world of superconductivity has highlighted both its remarkable achievements and the formidable challenges that lie ahead. Superconductivity, a phenomenon that eliminates electrical resistance and expels magnetic fields from the interior of a material, has the potential to revolutionize numerous technologies, including magnetic resonance imaging (MRI), levitated trains, and lossless power transmission. However, the quest to harness this phenomenon for practical applications faces significant hurdles, primarily due to the low temperatures required to achieve superconductivity in known materials.

The discovery of high-temperature superconductors, such as cuprates and iron-based superconductors, has shifted the landscape, pushing the boundaries closer to the coveted goal of room-temperature superconductivity. These materials, with their complex crystalline structures and unconventional pairing mechanisms, have challenged our understanding of superconductivity and opened new avenues for research. Theoretical models, including the BCS theory and its extensions, provide a framework for understanding superconductivity in conventional materials but fall short of fully explaining the properties of high-temperature superconductors. This gap in understanding underscores the need for a unified theory that can encompass both conventional and unconventional superconductors.

The technological implications of superconductivity are vast. In the realm of power transmission, superconducting cables could dramatically reduce energy losses, making electrical grids more efficient and sustainable. In transportation, superconducting magnets could enable faster and more efficient trains. In the field of computing, superconducting circuits could pave the way for quantum computers with unprecedented processing power. However, the widespread adoption of these technologies is contingent upon overcoming the challenges of cooling and material stability.

Research into new materials, such as hydrogen-rich compounds and novel alloys, continues to offer hope for achieving higher critical temperatures. Advances in materials science, combined with a deeper theoretical understanding of superconductivity, may eventually lead to the development of superconductors that operate at or near room temperature. Such a breakthrough would have profound implications for technology, energy, and transportation, marking a new era in the application of quantum phenomena.

In conclusion, superconductivity remains one of the most intriguing and promising areas of physics and materials science. Its potential to transform technology and society is immense, but realizing this potential requires overcoming significant scientific and engineering challenges. As research continues to push the boundaries of what is possible, the dream of practical, room-temperature superconductors moves ever closer to reality. The journey of discovery and innovation in superconductivity is far from over, and the coming years promise to be filled with exciting developments and breakthroughs.

References

1. H.K. Onnes, *The Superconductivity of Mercury* (Communications from the Physical Laboratory of the University, Leiden, 1911), pp.122–124
2. D. Breyel, T.L. Schmidt, A. Komnik, Rydberg crystallization detection by statistical means. Phys. Rev. A **86**(2) (2012)
3. M. Rini, Explaining mercury's superconductivity, 111 years later. Physics **15** (2022)
4. D. Ansermet, *Superconductivity: History and Motivations* (Springer Singapore, 2018), pp. 1–4
5. Y. Zhang, C.H. Wong, J. Shen, S.T. Sze, B. Zhang, H. Zhang, Y. Dong, H. Xu, Z. Yan, Y. Li, X. Hu, R. Lortz, Dramatic enhancement of superconductivity in single-crystalline nanowire arrays of Sn. Sci. Rep. **6**(1) (2016)
6. J. Baixeras, Les supraconducteurs: Applications à l'électronique et à l'électrotechnique. Sciences et Techniques de l'Ingénieur (1998)
7. R.V. Vovk, N.R. Vovk, G.Y. Khadzhai, I.L. Goulatis, A. Chroneos, Effect of high pressure on the electrical resistivity of optimally doped $YBa_2Cu_3O_{7-\delta}$ single crystals with unidirectional planar defects. Physica B: Condens. Matter **422**, 33–35 (2013)
8. R.P. Huebener, *The Meissner–Ochsenfeld Effect* (CRC Press, 2022), pp. 53–55
9. D. Hykel, Microscopie à micro-squid : étude de la coexistence de la supraconductivité et du ferromagnétisme dans le composé UCoGe. Ph.D. thesis, Grenoble (2011). Thèse de doctorat dirigée par Hasselbach, Klaus Physique
10. F. London, H. London, The electromagnetic equations of the supraconductor. Proc. R. Soc. Lond. Ser. A - Math. Phys. Sci. **149**(866), 71–88 (1935)
11. J.E. Hirsch, The origin of the meissner effect in new and old superconductors. Physica Scripta **85**(3), 035704 (2012)
12. M. Maier, Bcs theory in the weak magnetic field regime for systems with nonzero flux and exponential estimates on the adiabatic theorem in extended quantum lattice systems (2022)
13. R. Travaglino, A. Zaccone, Extended analytical bcs theory of superconductivity in thin films. J. Appl. Phys. **133**(3) (2023)

14. J. Bardeen, L.N. Cooper, J.R. Schrieffer, Theory of superconductivity. Phys. Rev. **108**(5), 1175–1204 (1957)
15. P.J. Ray, Master's thesis: structural investigation of La(2-x)Sr(x)CuO(4+y) - following staging as a function of temperature. figshare.com (2016)
16. J.F. Cochran, D.E. Mapother, Superconducting transition in aluminum. Phys. Rev. **111**(1), 132–142 (1958)
17. B.T. Matthias, T.H. Geballe, V.B. Compton, Superconductivity. Rev. Modern Phys. **35**(1), 1–22 (1963)
18. J. Eisenstein, Superconducting elements. Rev. Modern Phys. **26**(3), 277–291 (1954)
19. B.T. Matthias, T.H. Geballe, S. Geller, E. Corenzwit, Superconductivity of Nb_3Sn. Phys. Rev. **95**(6), 1435 (1954)
20. G.-I. Oya, E.J. Saur, Preparation of Nb_3Ge films by chemical transport reaction and their critical properties. J. Low Temp. Phys. **34**(5–6), 569–583 (1979)
21. Y. Kubo, K. Yoshizaki, Y. Hashimoto, Superconducting properties of chevrel-phase pbmo6s8 wires. Synth. Metals **18**(1–3), 851–856 (1987)
22. J.G. Bednorz, K.A. Müller, Possible high T_c superconductivity in the Ba-La-Cu-O system. Zeitschrift fur Physik B Condensed Matter **64**(2), 189–193 (1986)
23. M.K. Wu, J.R. Ashburn, C.J. Torng, P.H. Hor, R.L. Meng, L. Gao, Z.J. Huang, Y.Q. Wang, C.W. Chu, Superconductivity at 93 k in a new mixed-phase y-ba-cu-o compound system at ambient pressure. Phys. Rev. Lett. **58**(9), 908–910 (1987)
24. H. Maeda, Y. Tanaka, M. Fukutomi, T. Asano, A new high-tc oxide superconductor without a rare earth element. Jpn. J. Appl. Phys. **27**(2A), L209 (1988)
25. M.P. Siegal, E.L. Venturini, B. Morosin, T.L. Aselage, Synthesis and properties of Tl-Ba-Ca-Cu-O superconductors. J. Mater. Res. **12**(11), 2825–2854 (1997)
26. J. Nagamatsu, N. Nakagawa, T. Muranaka, Y. Zenitani, J. Akimitsu, Superconductivity at 39k in magnesium diboride. Nature **410**(6824), 63–64 (2001)
27. D. van Delft, The liquefaction of helium. Europhys. News **39**(6), 23–25 (2008)
28. C. Yao, *Fabrication and Properties of High-Performance 122-Type Iron-Based Superconducting Wires and Tapes* (Springer Nature Singapore, 2022), pp. 1–19
29. D. van Delft, History and significance of the discovery of superconductivity by Kamerlingh Onnes in 1911. Phys. C: Supercond. **479**, 30–35 (2012)
30. L. Kantorovich, *Superconductivity* (Springer, Netherlands, 2004), pp. 367–419
31. R. Giannetta, A. Carrington, R. Prozorov, London penetration depth measurements using tunnel diode resonators. J. Low Temp. Phys. **208**(1–2), 119–146 (2021)
32. S. Ghimire, K.R. Joshi, A. Datta, A. Goerdt, M.A. Tanatar, D. Schlagel, M.J. Kramer, J. Marshall, C.J. Copas, J.Y. Mutus, A. Romanenko, A. Grassellino, R. Prozorov, Quasiparticle spectroscopy in technologically-relevant niobium using London penetration depth measurements (2023)
33. S.G. Castillo-López, R. Esquivel-Sirvent, G. Pirruccio, C. Villarreal, The Abrikosov vortex structure revealed through near-field radiative heat exchange (2023)
34. S. Sachdev, *4 - Bardeen–Cooper–Schrieffer Theory of Superconductivity* (Cambridge University Press, 2023), pp. 43–52
35. E. Langmann, C. Triola, Universal and nonuniversal features of bardeen-cooper-schrieffer theory with finite-range interactions. Phys. Rev. B **108**, 104503 (2023)
36. B.W. Petley, The Josephson effects. Contemp. Phys. **10**(2), 139–158 (1969)
37. P. Mangin, R. Kahn, *The Josephson Effect* (Springer International Publishing, 2016), pp. 243–288
38. R.A. Ferrell, *The Josephson Effect* (Springer, New York, 1990), pp. 60–83
39. W.H.H. Granicher, K. Alex Müller and J. Georg Bednorz as graduate students at eth zürich. Ferroelectrics **89**(1), iii–ix (1989)
40. A. Khurana, Bednorz and Müller win nobel prize for new superconducting materials. Phys. Today **40**(12), 17–19 (1987)
41. A.M. Taheri, H. Ebrahimnezhad, M.H. Sedaaghi, Prediction of the critical temperature of superconducting materials using image regression and ensemble deep learning. Mater. Today Commun. **33**, 104743 (2022)

42. J.D. Boyes, N.H. Clark, Technologies for energy storage. Flywheels and super conducting magnetic energy storage, in *2000 Power Engineering Society Summer Meeting (Cat. No.00CH37134)*, PESS-00 (IEEE, 2000)
43. J. Rogers, Magnetic energy storage. IEEE Trans. Magn. **17**(1), 330–335 (1981)
44. D.R. Nayak, R. Dash, B. Majhi, Brain mr image classification using two-dimensional discrete wavelet transform and adaboost with random forests. Neurocomputing **177**, 188–197 (2016)
45. H. Maeda, Y. Tanaka, M. Fukutomi, T. Asano, A new high-T_c oxide superconductor without a rare earth element. Jpn. J. Appl. Phys. **27**(2A), L209 (1988)
46. Z.Z. Sheng, A.M. Hermann, Superconductivity in the rare-earth-free Tl-Ba-Cu-O system above liquid-nitrogen temperature. Nature **332**(6159), 55–58 (1988)
47. W. Meissner, R. Ochsenfeld, Ein neuer effekt bei eintritt der supraleitfahigkeit. Die Naturwissenschaften **21**(44), 787–788 (1933)
48. S.F. Saipuddin, A. Hashim, N.E. Suhaimi, *Type I and Type II Superconductivity* (Springer Nature Singapore, 2022), pp. 123–146
49. P.H. Mangin, Supraconductivité: un condensét de physique. Séminaire IFR Matériaux, École des Mines de Nancy (2003). Presentation
50. R.G. Sharma, *A Review of Theories of Superconductivity* (Springer International Publishing, 2015), pp. 109–133
51. S. Sachdev, *Landau–Ginzburg Theoryfrom Part I - Background* (Cambridge University Press, 2023), pp. 59–65
52. D. Ter Haar (ed.), *30 - On the Theory of Superconductivity* (Pergamon, 1965), pp. 217–225
53. A.C. Rose-Innes, E.H. Rhoderick, *The Mixed State* (Elsevier, 1978), pp. 183–201
54. A. Khanukov, I. Mangel, S. Wissberg, A. Keren, B. Kalisky, Mixed superconducting state without applied magnetic field. Phys. Rev. B **106**(14) (2022)
55. A.A. Abrikosov, L.M. Falicov, Fundamentals of the theory of metals. Phys. Today **43**(4), 73–76 (1990)
56. B.B. Goodman, The magnetic behavior of superconductors of negative surface energy. IBM J. Res. Dev. **6**(1), 63–67 (1962)
57. C. Kittel, J. Wiley, *Physique de l'état Solid* (Dunod, 2019)
58. T. B.P.G. de Gennes, P.A. Pincus, Superconductors: superconductivity of metals and alloys. Science **154**(3757), 288 (1966)
59. B.D. Josephson, Possible new effects in superconductive tunnelling. Phys. Lett. **1**(7), 251–253 (1962)
60. J. Gallop, A.I. Braginski, *Introduction to Section H3: Josephson Junction Devices* (CRC Press, 2022), pp. 640–641
61. J.G. Bednorz, K.A. Müller, Possible high T_c superconductivity in the Ba-La-Cu-O system. Zeitschrift fur Physik B Condensed Matter **64**, 189–193 (1986)
62. M.K. Wu, J.R. Ashburn, C.J. Torng, P.H. Hor, R.L. Meng, L. Gao, Z.J. Huang, Y.Q. Wang, C.W. Chu, Superconductivity at 93 K in a new mixed-phase Y-Ba-Cu-O compound system at ambient pressure. Phys. Rev. Lett. **58**(9), 908–910 (1987)
63. Z.Z. Sheng, A.M. Hermann, Bulk superconductivity at 120 K in the Tl-Ca/Ba-Cu-O system. Nature **332**(6160), 138–139 (1988)
64. C.W. Chu, L. Gao, F. Chen, Z.J. Huang, R.L. Meng, Y.Y. Xue, Superconductivity above 150 K in $HgBa_2Ca_2Cu_3O_{8+\delta}$ at high pressures. Nature **365**(6444), 323–325 (1993)
65. L. Gao, Y.Y. Xue, F. Chen, Q. Xiong, R.L. Meng, D. Ramirez, C.W. Chu, J.H. Eggert, H.K. Mao, Superconductivity up to 164 K in $HgBa_2Ca_{m-1}Cu_mO_{2m+2+\delta}$ (m=1, 2, and 3) under quasihydrostatic pressures. Phys. Rev. B **50**(6), 4260–4263 (1994)
66. Y. Kamihara, T. Watanabe, M. Hirano, H. Hosono, Iron-based layered superconductor $LaO_{1-x}F_x$ FeAs (x = 0.05 - 0.12) with T_c = 26 k. J. Amer. Chem. Soc. **130**(11), 3296–3297 (2008)
67. H. Takahashi, K. Igawa, K. Arii, Y. Kamihara, M. Hirano, H. Hosono, Superconductivity at 43 K in an iron-based layered compound $LaO_{1-x}F_x$FeAs. Nature **453**(7193), 376–378 (2008)
68. Inamuddin, R. Boddula, M.M. Rahman, A.M. Asiri (eds.), *Green Sustainable Process for Chemical and Environmental Engineering and Science* (Elsevier Science Publishing, Philadelphia, PA, 2021)

69. A. Mourachkine, *Room-temperature Superconductivity* (Cambridge International Science Publishing, Cambridge, England, 2004)

70. R.J. Cava, B. Batlogg, K.M. Rabe, E.A. Rietman, P.K. Gallagher, L.W. Rupp, Structural anomalies at the disappearance of superconductivity in $Ba_2YCu_3O_{7-\delta}$: evidence for charge transfer from chains to planes. Physica C: Supercond. **156**(4), 523–527 (1988)

71. Y. Liu, F. Xue, X.-F. Gou, Role of grain boundary networks in vortex motion in superconducting films. Chin. Phys. B **32**(12), 127401 (2023)

72. P. Lucignano, D. Stornaiuolo, F. Tafuri, B.L. Altshuler, A. Tagliacozzo, Evidence for a minigap in YBCO grain boundary Josephson junctions. Phys. Rev. Lett. **105**(14) (2010)

73. J.-F. Seaux, Conception, optimisation et test de dispositifs intégrant des matériaux en couche mince supraconducteurs ou ferroélectriques pour des applications de filtrage dans le domaine spatial. Ph.D. thesis, PHDTHESIS, 2005. Thèse de doctorat dirigée par Madrangeas, Valérie Électronique des hautes fréquences et optoélectronique. Communication optiques et micro-ondes Limoges (2005)

74. G.J. McIntyre, A. Renault, G. Collin, Domain and crystal structure of superconducting $Ba_2YCu_3O_{8-\delta}$ at 40 and 100 K by single-crystal neutron diffraction. Phys. Rev. B **37**(10), 5148–5157 (1988)

75. R.K. Patel, K. Patra, S.K. Ojha, S. Kumar, S. Sarkar, J.W. Freeland, J.W. Kim, P.J. Ryan, P. Mahadevan, S. Middey, Hole doping in a negative charge transfer insulator (2022)

76. K.H. Sarwa, B. Tan, K. Parendo, Y.-H. Lin, A.M. Goldman, Electrostatic modification of the conductive properties of amorphous bi ultrathin films. Phys. C: Supercond. **468**(4), 299–303 (2008)

77. B. Pignon, Contribution à l'étude de l'influence du dopage sur les propriétés électroniques des cuprates supraconducteurs. Ph.D. thesis, PHDTHESIS, 2005. Thèse de doctorat dirigée par Ammor, Larbi Physique Tours (2005)

78. Chi Ho Wong and Rolf Lortz, Preliminary T_c calculations for iron-based superconductivity in NaFeAs, LiFeAs, FeSe and nanostructured $FeSe/SrTiO_3$ superconductors. Materials **16**(13), 4674 (2023)

79. H. Hosono, H. Fukuyama, H. Akai, Iron-based superconductors are entering a new stage: actors are ready. Solid State Commun. **152**(8), 631 (2012). Special Issue on Iron-based Superconductors

80. Z.-A. Ren, G.-C. Che, X.-L. Dong, J. Yang, L. Wei, W. Yi, X.-L. Shen, Z.-C. Li, L.-L. Sun, F. Zhou, Z.-X. Zhao, Superconductivity and phase diagram in iron-based arsenic-oxides $ReFeAsO_{1-\delta}$ (re = rare-earth metal) without fluorine doping. EPL (Europhys. Lett.) **83**(1), 17002 (2008)

81. G. Wu, Y.L. Xie, H. Chen, M. Zhong, R.H. Liu, B.C. Shi, Q.J. Li, X.F. Wang, T. Wu, Y.J. Yan, J.J. Ying, X.H. Chen, Superconductivity at 56 K in samarium-doped srfeasf. J. Phys.: Condens. Matter **21**(14), 142203 (2009)

82. M. Rotter, M. Tegel, D. Johrendt, Superconductivity at 38 K in the iron arsenide $(Ba_{1-x}K_x)Fe_2As_2$. Phys. Rev. Lett. **101**(10) (2008)

83. K. Sasmal, B. Lv, B. Lorenz, A.M. Guloy, F. Chen, Y.-Y. Xue, C.-W. Chu, Superconducting fe-based compounds $(A_{1-x}Sr_x)Fe_2As_2$ with a = K and cs with transition temperatures up to 37 k. Phys. Rev. Lett. **101**(10) (2008)

84. C.W. Chu, F. Chen, M. Gooch, A.M. Guloy, B. Lorenz, B. Lv, K. Sasmal, Z.J. Tang, J.H. Tapp, Y.Y. Xue, The synthesis and characterization of lifeas and NaFeAs. Phys. C: Supercond. **469**(9–12), 326–331 (2009)

85. M.J. Pitcher, D.R. Parker, P. Adamson, S.J.C. Herkelrath, A.T. Boothroyd, R.M. Ibberson, M. Brunelli, S.J. Clarke, Structure and superconductivity of LiFeAs. Chem. Commun. **45**, 5918–5920 (2008)

86. J.H. Tapp, Z. Tang, B. Lv, K. Sasmal, B. Lorenz, P.C.W. Chu, A.M. Guloy, LiFeAs: an intrinsic feas-based superconductor with T_c = 18 k. Phys. Rev. B **78**(6) (2008)

87. D.R. Parker, M.J. Pitcher, P.J. Baker, I. Franke, T. Lancaster, S.J. Blundell, S.J. Clarke, Structure, antiferromagnetism and superconductivity of the layered iron arsenide NaFeAs. Chem. Commun. **16**, 2189 (2009)

88. Y. Zhang, X. Xiaojie, Predicting doped mgb2 superconductor critical temperature from lattice parameters using gaussian process regression. Phys. C: Supercond. Appl. **573**, 1353633 (2020)
89. D. Tripathi, T.K. Dey, Effect of (Bi, Pb)-2223 addition on thermal transport of superconducting MgB_2 pellets. J. Alloys Comp. **618**, 56–63 (2015)
90. A.F. Hebard, M.J. Rosseinsky, R.C. Haddon, D.W. Murphy, S.H. Glarum, T.T.M. Palstra, A.P. Ramirez, A.R. Kortan, Superconductivity at 18 K in potassium-doped C60. Nature **350**(6319), 600–601 (1991)
91. A.Y. Ganin, Y. Takabayashi, Y.Z. Khimyak, S. Margadonna, A. Tamai, M.J. Rosseinsky, K. Prassides, Bulk superconductivity at 38 K in a molecular system. Nat. Mater. **7**(5), 367–371 (2008)
92. L.-N. Zong, R.-S. Wang, D. Peng, X.-J. Chen, Robust superconductivity near constant temperature in rubidium-doped C_{60} (2022)
93. R.C. da Silva, C.C. Bastos, A.C. Pavão, High-T_c fullerides. Phys. C: Supercond. Appl. **561**, 13–17 (2019)

Chapter 2
High Critical Temperature Superconducting Oxides of the YBCO System

2.1 HCTO: High Critical Temperature Superconducting Oxides

2.1.1 Presentation of Materials

Most materials used in superconductivity studies belong to the family of copper oxides $RBa_2Cu_3O_{7-\delta}$ where R is a rare earth element. The majority of these compounds are perovskite-type ABO_3 oxygen-deficient superconductors with high critical temperatures. They possess a layered structure along the c-axis, composed of CuO_2 planes determining the superconducting character of these materials and Cu-O chains acting as charge reservoirs for the CuO_2 planes. The oxygenation state of the Cu-O chains determines the structure of these compounds as well as their anisotropy in the (a, b) plane.

2.1.2 Perovskite Structure

Numerous studies, including those by Hazen et al. [1], have shown that high critical temperature superconductors belong to the "perovskite" crystallographic family [2–4]. The ideal perovskite structure of highest symmetry is a cubic symmetry structure with the chemical formula ABX_3 (Fig. 2.1).

In this structure, the X atom is a non-metallic anion forming a network of octahedra connected at their vertices. A and B are metallic cations. The B element occupies the center of each octahedron while the A element, with a larger atomic radius, is located in the middle of eight octahedra (Fig. 2.2).

K. Khallouq, *Exploring High-Temperature Superconductivity in the YBCO System*, SpringerBriefs in Materials, https://doi.org/10.1007/978-3-031-66238-6_2

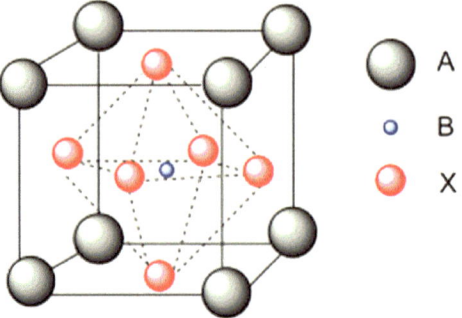

Fig. 2.1 Perovskite structure ABX₃

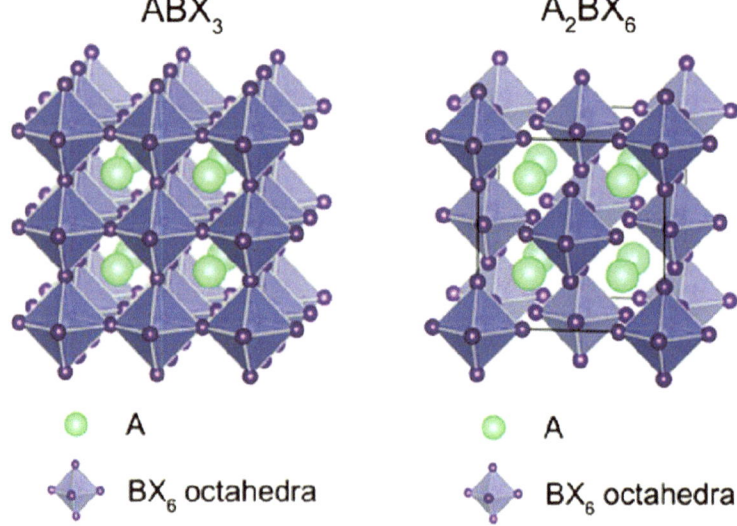

Fig. 2.2 Crystal structures of ABX₃ (left) and A₂BX₆ (right) [5]

2.1.3 YBa₂Cu₃O₇₋δ Structure

The compound $YBa_2Cu_3O_{7-\delta}$, like most superconducting cuprates, is part of a family of oxides (with δ between 0 and 1) and is formed from three oxygen-deficient perovskite units (B = Cu, A = Y for the middle cube, and A = Ba for the other two cubes) (Fig. 2.3). Its structure is composed of CuO_2 planes in the a-b direction and Cu-O chains along the b-axis.

For $0 \leq \delta \leq 0.65$, the $YBa_2CuO_{7-\delta}$ structure becomes superconducting, and its elementary cell is orthorhombic. The physical properties of $YBa_2CuO_{7-\delta}$ are very sensitive to its oxygen content $(7 - \delta)$: the critical temperature varies with the δ rate, showing a maximum $T_c = 92$ K for $\delta = 0$. As δ decreases, oxygen atoms are

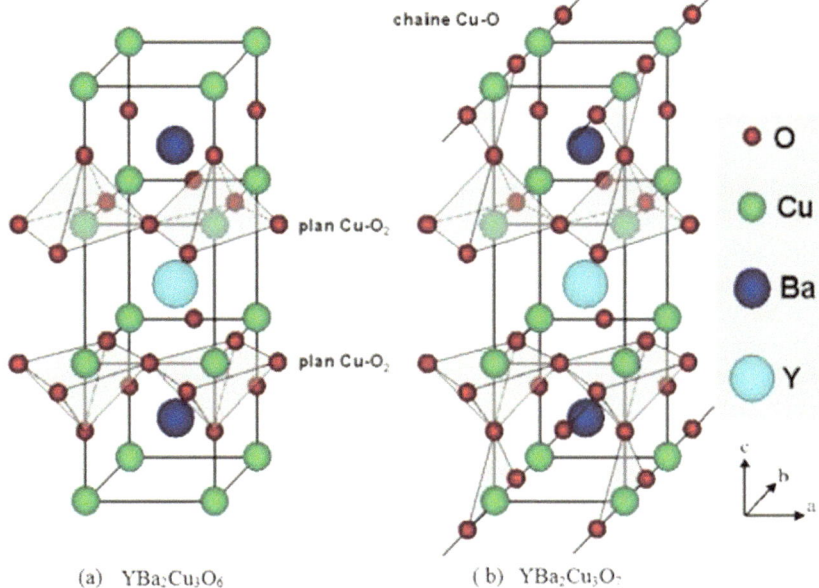

chaine Cu-O

plan Cu-O₂

plan Cu-O₂

O

Cu

Ba

Y

(a) YBa₂Cu₃O₆   (b) YBa₂Cu₃O₇

Fig. 2.3 Crystallographic structure of compounds **a** YBa_2CuO_6 with a tetragonal structure ($a_0 = b_0 = 3.86$ Å, $c_0 = 11.7$ Å), **b** YBa_2CuO_7 with an orthorhombic structure ($a_0 = 3.823$ Å, $b_0 = 3.885$ Å, $c_0 = 11.7$ Å) [6]

inserted along the b-axis and form Cu-O chains. The formation of these chains causes a distortion of the cell and induces hole doping in the CuO_2 planes and electron doping in the Cu-O chains. The presence of oxygen atoms in these chains is essential for superconductivity. For $\delta > 0.65$, the structure is antiferromagnetic, and its elementary cell is tetragonal. In both cases ($\delta = 1$ and $\delta = 0$), there are no O^{2-} ions in the planes perpendicular to the c-axis containing Y^{3+} ions. However, all the oxygen ions are present in the planes containing Ba^{2+} ions. There are two types of copper ions:

- Cu_1 are located on a site with four oxygens as nearest neighbors. These are the Cu-O chains along the b-axis. Along the a-axis, there is a lack of oxygen between the copper atoms. The oxygen ions in the Cu-O chains attract electrons provided by the CuO_2 planes, thus doping these planes with holes, as demonstrated by Rietschel et al. through electronic spectroscopy measurements [7].
- Cu_2 are located in the middle of the base of a tetragonal pyramid with five oxygens as nearest neighbors. The sixth place of the oxygen octahedron is not occupied. The bases of these pyramids form the CuO_2 layers (which will play the role of free carrier reservoirs that will dope these planes and make them superconducting).

2.1.4 CuO₂ Planes

There is a physical relationship between the critical temperature T_c and the presence of CuO_2 planes (ab planes), which is the common point to all superconducting cuprates. They are formed of divalent copper atoms (Cu^{2+}) and oxygen ions (O^{2-}) placed on a square lattice (Fig. 2.4). The elementary cell contains n layers of CuO_2 planes, each separated by yttrium (Y) atoms and framed by two blocks containing metals, rare earths, and oxygen which constitute positive charge reservoirs.

It is currently accepted that superconductivity appears at the level of CuO_2 planes. The "reservoir" blocks act as insulators, channeling the current in a specific direction.

The different families of HTSC are defined by the nature of the atoms used in the insulating blocks. The main ones are those based on bismuth (Bi), thallium (Tl), and mercury (Hg), as well as YBCO. Each family contains a large number of superconducting oxides since the number of CuO_2 planes plays a fundamental role, as does the stoichiometry of the compound [8].

2.1.5 Other Structures

2.1.5.1 La₂₋ₓAₓCuO₄ Structure

The $La_{2-x}A_xCuO_4$ compounds, with A being an alkaline earth element (A = Ba, Sr, Ca), are part of the "214" group. They are considered prototypes because their

Fig. 2.4 CuO₂ plane formed of CuO₄ squares sharing an oxygen at each corner

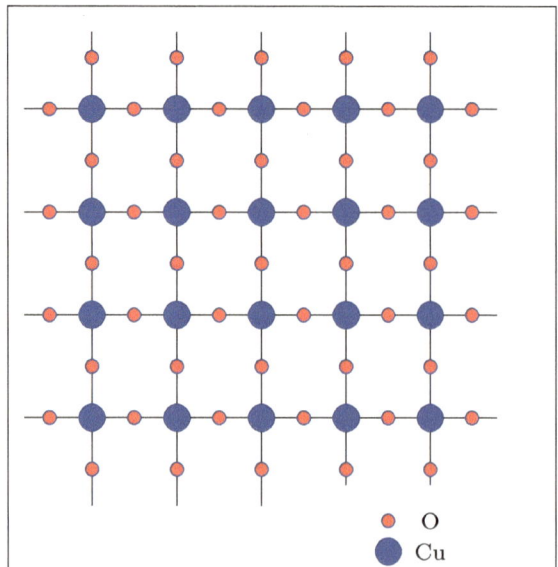

structure is simple: two CuO_2 layers per crystallographic unit "sandwiched" between LaO layers, making these materials part of the cuprate family. Substitution at the La site is possible. $La_{2-x}A_xCuO_4$ exhibits the highest T_c around a critical value of x (x = 0.15 for Sr and 0.2 for Ba with T_c around 30 K and 35 K, respectively) [9, 10].

2.1.5.2 Bi-Based Cuprate Structure

$Bi_2Sr_2Ca_{n-1}Cu_nO_{2n+4+x}$ is the general formula for the bismuth-based superconductor family (BSCCO). The oxygen stoichiometry x plays a fundamental role in the superconducting properties [11–14], and is difficult to control. The formula consists of three main phases. The first phase is the compound $Bi_2Sr_2CuO_{6+x}$ (Bi2201) with $n = 1$, synthesized by Professor B. Raveau's team in 1987 [15]. A year later, Maeda et al. [16] synthesized the second phase, the compound $Bi_2Sr_2CaCu_2O_{8+x}$ (Bi2212) with $n = 2$ at a $T_c = 58$ K. Later, the compound $Bi_2Sr_2Ca_2Cu_3O_{10+x}$ (Bi2223) with $n = 3$ was synthesized by Tarascon et al. [17], achieving a $T_c = 110$ K. These phases are of orthorhombic symmetry.

The crystallographic parameters of the Bi2201 phase's elementary cell are: $a = 5.371$ Å, $b = 5.372$ Å, and $c = 24.59$ Å [15] with a single perovskite layer formed of CuO_6 octahedra (Fig. 2.5a). The parameters of the Bi2212 phase's elementary cell are: $a = 5.395$ Å, b = 5.39 Å, and $c = 30.65$ Å [18] with sheets of CuO_5 pyramids separated by a calcium plane (Fig. 2.5b). The Bi2223 phase has an elementary cell with parameters: $a \approx b = 5.4$ Å and $c = 37$ Å [19]. The structure contains a CuO_4 square plane framed by two calcium planes and two layers of CuO_5 pyramids (Fig. 2.5). This class of superconductors does not contain harmful elements like those based on mercury or thallium. Moreover, in terms of implementation, bismuth compounds show great inertness against corrosive agents, especially H_2O and CO_2, and have the advantage of oxygen composition stability, for instance, some of these superconducting phases do not lose oxygen when annealed at ≈ 850 °C, unlike the Y123 compound.

2.1.5.3 Thallium-Based Superconducting Compound Structure

In October 1987, Sheng and Hermann discovered the thallium-based cuprate (Tl-Ba-Cu-O) system in compounds $Tl_2Ba_2Cu_3O_{8+x}$ and $TlBa_2Cu_3O_{5.5+x}$ with a T_c above 80 K [21]. They added Ca to their samples and discovered superconductivity above 100 K in the Tl-Ba-Ca-Cu-O system [22, 23] (Fig. 2.6). Later, Parkin et al. [24] modified the processing conditions and achieved a superconducting phase with a critical temperature reaching 125 K in the compound $Tl_2Ba_2Ca_2Cu_3O_{10}$.

Thallium-based compounds are numerous [25–27]. They can be described by the general formula: $Tl_mBa_2Ca_{n-1}Cu_nO_{(3/2)m+2n+2}$ where $m = 1$ or 2 and $n = 1$, 2, 3, or 4. n represents the number of CuO_2 planes, separated by a calcium plane for $n > 1$ (Fig. 2.6), and m corresponds to the number of consecutive layers of TlO planes. The critical temperature increases with the number of CuO_2 planes, and it is

Fig. 2.5 Crystallographic structure of three compounds (from left to right) Bi2201, Bi2212, and Bi2223 [20]

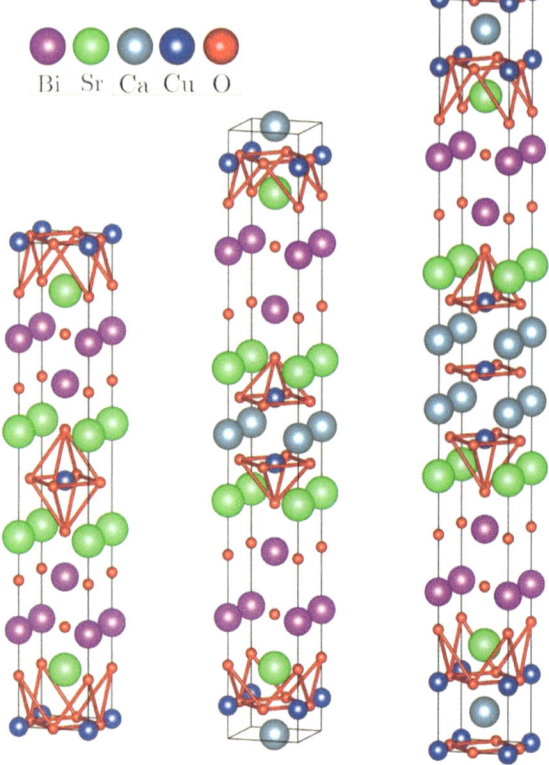

Bi Sr Ca Cu O

15–20 K higher for compounds of the Tl(2) family compared to those of the Tl(1) family, corresponding, respectively, to $m = 2$ and $m = 1$, for the same n. For the Tl(2) compounds, Shimakawa et al. [28] showed that as the number of CuO_2 planes increased, the Cu-O(2) distance decreased (O(2): the apical site of copper octahedra or pyramids). This phenomenon seems to indicate that T_c is strongly correlated with the structure.

2.1.5.4 Mercury-Based Cuprate Superconductors

All superconducting cuprates possess layered copper oxide CuO_2 structures alternating with other structural blocks that act as charge reservoirs. In the majority of cuprates exhibiting the highest critical temperatures, the reservoir blocks consist of a monolayer of thallium oxide (TlO) or a double layer $(TlO)_2$ or $(BiO)_2$ [29, 30]. The mercury-based cuprate family $HgBa_2RCu_2O_{6+\delta}$ (Hg-1212), where R is a rare earth, was synthesized by Putilin et al. [31]. It has a structure similar to that of the superconducting thallium $TlBa_2CaCu2O_7$ (Tl-1212), featuring a TlO layer

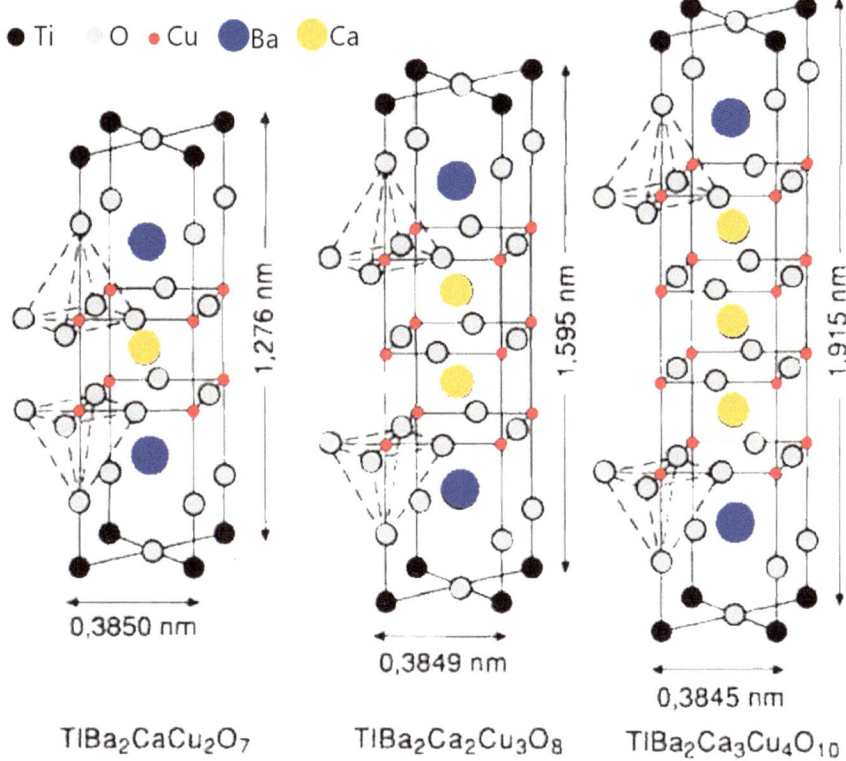

● Ti ○ O • Cu 🔵 Ba 🟡 Ca

$TlBa_2CaCu_2O_7$ $TlBa_2Ca_2Cu_3O_8$ $TlBa_2Ca_3Cu_4O_{10}$

Fig. 2.6 Crystallographic structure of three compounds (from left to right) Tl-1212, Tl-1223, and Tl-1234

and two CuO_2 layers per unit cell and a $T_c \approx 85$ K [32]. Despite the resemblance to Tl-1212, Hg-1212 was deemed non-superconducting (as the hole concentration in its phases might not be high enough to induce superconductivity). In September 1993, Putilin et al. [33] described the synthesis of the rare-earth-free superconducting compound $HgBa_2RCu_2O_{6+\delta}$ (Hg-1201), containing only one CuO_2 plane but exhibiting a T_c of 94 K ((a)). Its structure is similar to that of Tl-1201 ($T_c \leq 01$ K), but its T_c is much higher. The availability of such a high-T_c material with a single metallic oxide layer (HgO) may be significant for technological applications due to a reduced spacing between CuO_2 planes, leading to better superconducting properties in a magnetic field. Putilin et al. [34] replaced the trivalent rare earths with the divalent cation Ca^{2+} and achieved a superconducting transition temperature above 120 K in $HgBa_2RCaCu_2O_{6+\delta}$ (Fig. 2.7).

In April 1993, Schilling et al. [36] detected superconductivity at critical temperatures above 130 K in samples consisting of $HgBa_2RCaCu_2O_{6+\delta}$ (Hg-1212) (with two CuO_2 layers per cell (Fig. 3.7b)) and $HgBa_2Ca_2Cu_3O_{8+\delta}$ (Hg-1223) (with three CuO_2 layers per cell). The high T_c was attributed to the Hg-1223 compound.

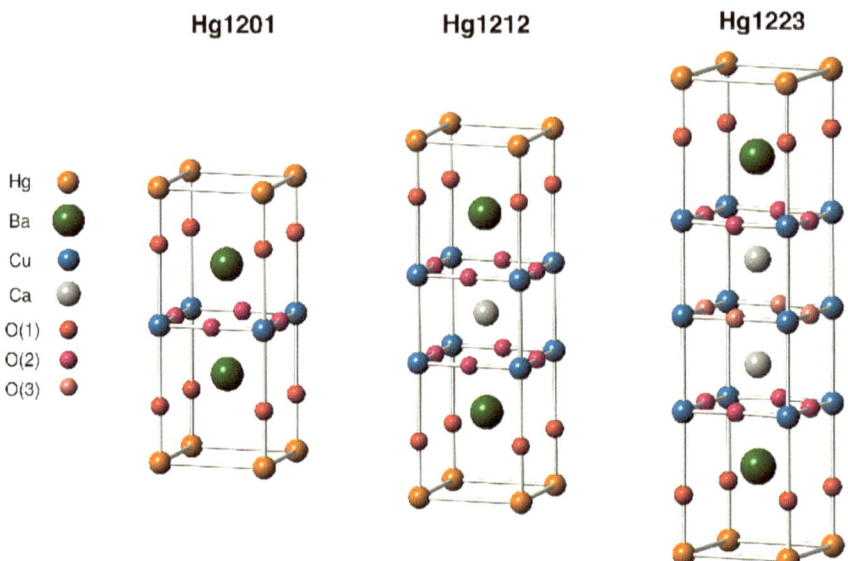

Fig. 2.7 Structural comparison of Hg-family cuprates [35]

2.2 Study of the YBCO Prototype

2.2.1 Crystal Structure

Discovered in 1987 by Chu and Wu, the compound $YBa_2Cu_3O_{6+\delta}$ is shown in Fig. 3.8. To date, it remains the most well known among the cuprates. It was the first to have a critical temperature ($T_c = 92$ K) exceeding the symbolic barrier of liquid nitrogen ($T = 77$ K) and is easier to synthesize [37]. Moreover, the oxygen content of these cuprates depends on their preparation conditions and can be controlled. The structure of this compound is a triple pseudo-perovskite. It consists of three cells, one centered on yttrium (Y) and the other two on barium (Ba) (Fig. 2.8). The cell displays five non-equivalent crystallographic oxygen sites noted O(i) ($i = 1$ to 5) and two distinct copper sites Cu(1) (Cu(1) belongs to Cu-O chains, parallel to the b-axis) and Cu(2) (Cu(2) belongs to CuO$_2$ planes), with Cu(1) having a coordination of 2 for $\delta = 0$ and 4 for $\delta = 1$, and Cu(2) always in coordination 5 (pyramidal environment).

The $YBa_2Cu_3O_7$ compound is orthorhombic, with parameters $a = 3.82$ Å, $b = 3.89$ Å and $c = 11.68$ Å. The unit cell is formed by alternating layers of CuO–BaO–CuO$_2$–Y–CuO$_2$–BaO–CuO. The Cu-O(4) planes consist of a set of parallel chains along the b-axis. The O(4) site accommodating oxygen can be either vacant or filled, allowing for a continuously variable stoichiometry from the compound $YBa_2Cu_3O_7$ to $YBa_2Cu_3O_6$. From oxidation $\delta = 0.4$ to $\delta = 1$, additional oxygens occupy the O(4) sites. They capture electrons due to their electronegative character, or, alternatively, release holes. These positively charged holes will be located on the orbitals of the

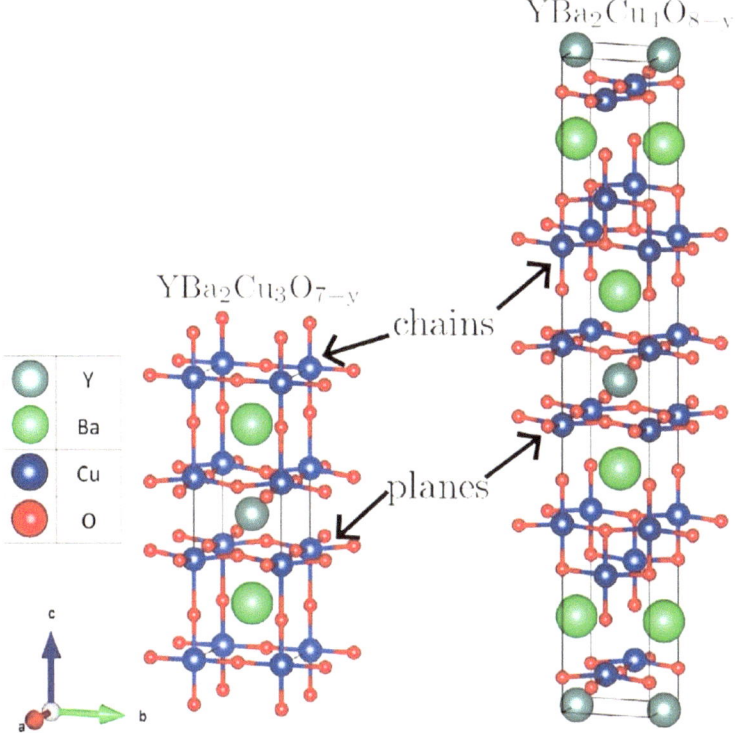

Fig. 2.8 Left: YBa$_2$Cu$_3$O$_{7-\delta}$ (Y-123) unit cell. Right: YBa$_2$Cu$_4$O$_{8-\delta}$ (Y-124) crystal structure [38]

oxygens in the CuO$_2$ planes. For $0 < \delta < 0.4$, additional oxygens are randomly placed on the O(4) or O(5) sites (the two lacunar sites of the Cu(1)-O planes).

2.2.2 Phase Diagram

The phase diagram corresponding to YBa$_2$Cu$_3$O$_{6+\delta}$ as a function of partial oxygen content ($0 \le \delta \le 1$) [39] is represented in Fig. 2.9. It is clear that as the oxygen content decreases, T_c diminishes until it becomes zero around $\delta = 0.4$. For an oxygen composition lower than 6.4, the compound becomes antiferromagnetic with a Néel temperature, T_N, around 400 K for YBa$_2$Cu$_3$O$_6$ [39, 40]. The transition between the superconducting and antiferromagnetic phases occurs when the compound shifts from the orthorhombic to the tetragonal structure. A T_c peak around 92 K is observed for $\delta = 0.95$, and the critical temperature drops to 89 K when δ varies from 0.95 to 1. The compound at the maximum doping $\delta = 1$ is thus slightly overdoped, in the sense that its T_c is lower than the maximum T_c.

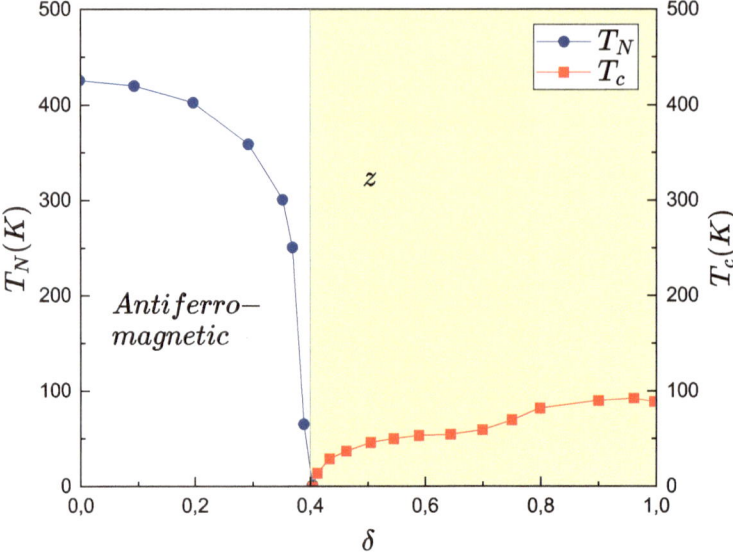

Fig. 2.9 Phase diagram of $YBa_2Cu_3O_{6+\delta}$ as a function of partial oxygen content z [41]

- For $0.0 < \delta < 0.4$: The compound is an insulating tetragonal and antiferromagnetic.
- For $0.4 < \delta < 1$: The compound is orthorhombic, metallic p-type, and becomes superconducting at low temperature ($T \leq 94\,K$).

2.2.2.1 Effect of Oxygen Stoichiometry on $YBa_2Cu_3O_{6+\delta}$

On the Critical Temperature: The oxygen stoichiometry conditions all properties of the compound $YBa_2Cu_3O_{6+\delta}$ (structural, electrical, and magnetic properties) [37, 42]. There are two possibilities for modulating the amount of oxygen in this type of material [43]:

- Either by replacing the cations in the reservoir layers with others of different valence (example: Ba^{2+} by La^{3+}).
- Or by adjusting the thermal treatments (temperature and partial oxygen pressure, as well as cooling rate).

Moreover, it's important to remember that the critical temperature T_c is very sensitive to the oxygen content z, as can be seen in Fig. 2.10. When the oxygen concentration of the compound $YBa_2Cu_3O_{6+\delta}$ is below approximately 6.4, it no longer exhibits superconductivity regardless of the temperature [44–46]. The composition of $YBa_2Cu_3O_7$ corresponds to a maximum value for oxygen stoichiometry. Conversely, sufficient deoxygenation (for example, heating under vacuum) leads to a

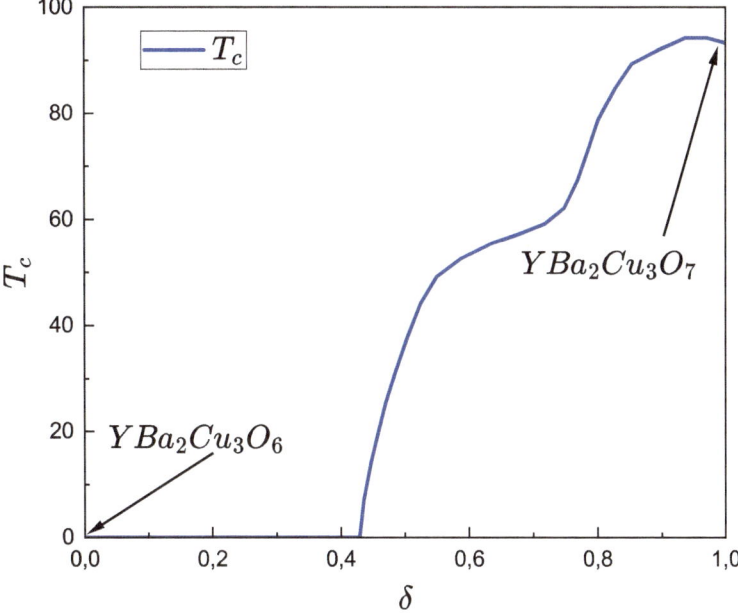

Fig. 2.10 Influence of oxygen content δ on the critical temperature of the compound $YBa_2Cu_3O_{6+\delta}$

phase even more deficient in oxygen, $YBa_2Cu_3O_6$. The structure of this phase is obtained by simply removing the O(4) oxygens from Fig. 2.7.

Thus, for $\delta = 1$, the insertion of oxygen at the O(4) site increases b, resulting in an orthorhombic structure. We have four oxygens, arranged in the plane $(\vec{b}, \vec{c})$, centered on the chain copper Cu(1) (Fig. 2.7). The formula for neutral charge is written: Y^{3+} $(Ba^{2+})_2 (Cu^{2+})_2(Cu^{3+}) (O^{2-})_7$.

Copper Cu(2) from the Cu(2)O$_2$ planes always has (for $\delta = 0$ or 1) a coordination 5 with a square-based pyramidal environment (2O(2), 2O(3), and 1O(1)).

For $\delta = 0$, the structure is tetragonal ($a = b$). The copper of the Cu(1) chains, between two Ba-O planes, has a linear coordination 2 along $\vec{c}$.

The critical temperature T_c of the compound $YBa_2Cu_3O_{6+\delta}$ depends on its oxygen content. It goes from 90 K (for $\delta = 1$) to about 40 K (for $\delta = 0.5$), with a maximum ($\approx$92 K) at z = 0.95 as shown in Fig. 2.10 [47]. Superconductivity is destroyed when $\delta \approx 0.45$.

On the Lattice Parameters: The oxygen content δ also significantly affects the crystal structure of the compound $YBa_2Cu_3O_{6+\delta}$, transitioning from an orthorhombic to a tetragonal system, resulting in the disappearance of superconductivity. When the oxygen concentration is equal to 6 ($\delta = 0$), the crystal structure of YBCO is tetragonal, and the a- and b-axes are equal [44–46]. However, the c-axis is much larger (Fig. 2.11). Gradually increasing the oxygen concentration reveals a length

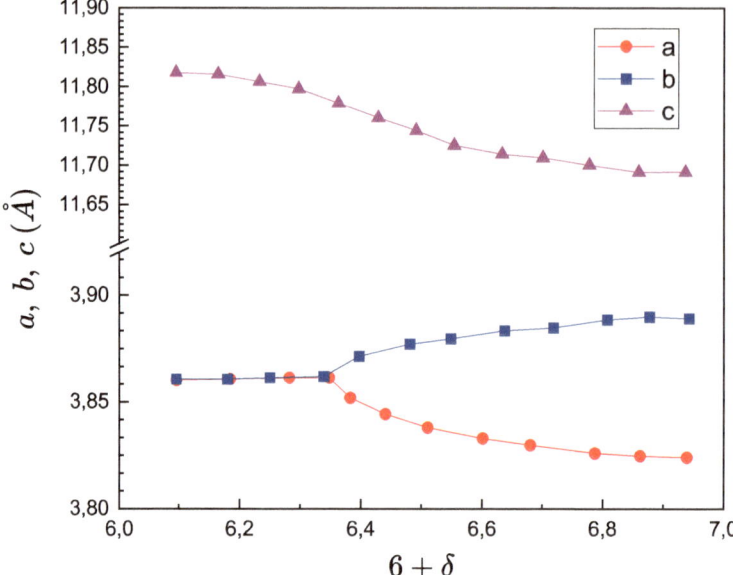

Fig. 2.11 Effect of oxygen deficiency on T_c of the compound $YBa_2Cu_3O_{6+\delta}$

difference between the a- and b-axes, becoming more pronounced. The crystal structure gradually transforms into an orthorhombic phase, which is the crystal phase in which YBCO is superconducting. Figure 2.10 shows that additional oxygens in the orthorhombic phase attach to the copper–oxygen Cu-O chains. Meanwhile, an increase in oxygen concentration further increases the electrostatic attraction between planes (charged with holes) and chains (charged with electrons), bringing the planes closer together and decreasing the c parameter.

2.2.2.2 Charge Transfer Model

As shown in Fig. 2.12 (variation of T_c, B.V.S (Cu_2), and B.V.S (Cu_1) as a function of oxygen content z), the main effect of deoxygenation is to progressively destroy the charge transfer from Cu-O chains to the conducting CuO_2 planes. The decrease in T_c from 90 to 60 K and the transition from the superconducting state to the semiconductor state correspond, respectively, to a decrease in positive charge (holes) in the $Cu(2)$-O_p planes of 0.03 and 0.05 per $Cu(2)$ atom. In the tetragonal phase B.V.S (Cu_2) $= 2 = const$ and B.V.S (Cu_1) $= f(z)$. As δ increases, oxygen O^{2-} fills the O(4) sites, and superconductivity appears for $\delta > 0.4$ and $T < T_c$. Hence the charge transfer model: the chains are charge reservoirs that feed the $Cu(2)$-O_p planes, where superconductivity occurs, via the apical oxygens O(1).

Moreover, recent studies have shown that the electronic structure of the compound $YBa_2Cu_3O_7$ [48] established that it's not so much the overall oxygen composition that

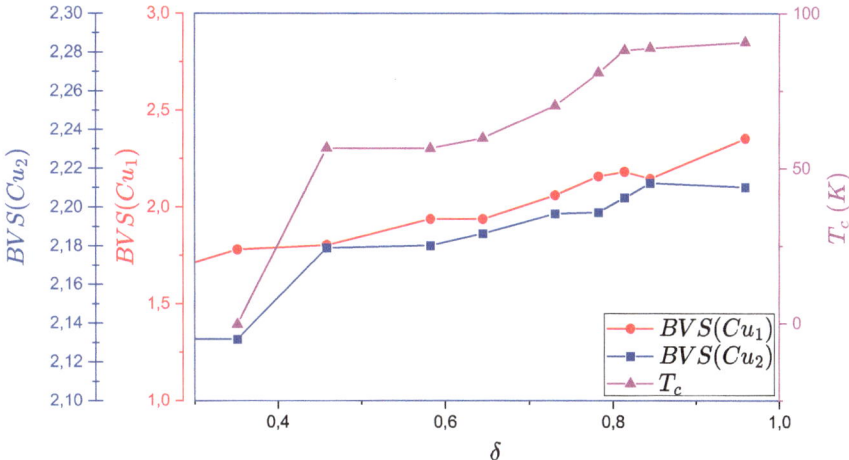

Fig. 2.12 Correlation between the number of carriers in the CuO planes, CuO chains, and the critical temperature for the compound $YBa_2Cu_3O_{6+\delta}$ with different oxygen contents

determines the material's electronic properties as the coordination of the copper in the $Cu(1)$ chains. When it is square planar, a band forms below the Fermi level capable of accepting electrons from the band associated with the CuO_2 planes. In this case, the oxide is superconducting, and its lattice is orthorhombic. Conversely, when oxygen atoms are equally distributed along the b- and a-axes (sites $O(4)$ and $O(5)$), this band disappears, and the Fermi level is at the tail of the conduction band. The material is insulating, and its lattice is tetragonal. Thus, the evolution of electronic properties through deoxygenation is linked to a modification of the material's band structure, consequently decreasing the charge transfer. For the compound $YBa_2Cu_3O_{6+\delta}$, three types of structures are distinguished, with different T_c depending on z [49]:

- For $\delta = 0$: The structure is tetragonal (T), both $O(4)$ and $O(5)$ sites are vacant.
- For $0.8 < \delta < 1$: The YBCO compound has a $T_c = 90$ K to 92 K, the $O(5)$ sites are always vacant, and the structure is orthorhombic I (ortho-I). The optimal T_c of 92 K is obtained for $\delta = 0.95$.
- For $0.45 < \delta < 0.6$: In which YBCO has a $T_c = 60$ K, half of the $O(4)$ sites and all $O(5)$ sites are vacant. The structure is orthorhombic II (ortho-II).

The oxygen concentration dependence of these phases is also visible in Fig. 2.13. Orthorhombicity ϵ and the site order parameter SOP are defined by [b − a/b + a] and [n(O_4) − n(O_5)/n(O_4) + n(O_5)], respectively, where n(O_4) and n(O_5) are the numbers of oxygens at sites $O(4)$ and $O(5)$ in the basal plane. The former can be considered a macroscopic order parameter, while the latter is a microscopic order parameter [49–51]. Orthorhombicity ϵ increases with SOP (Fig. 2.14), from 0 to almost $\epsilon = 10^{-2}$ (SOP = 1), obtained for oxygen order in the chain plane in an orthorhombic structure (ortho-I).

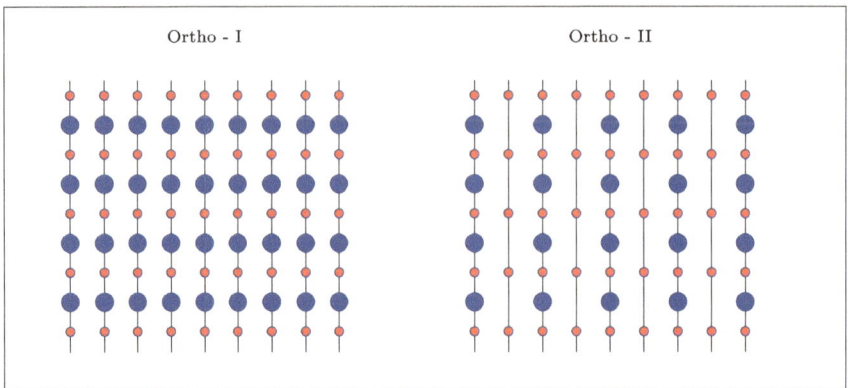

Fig. 2.13 Diagram showing the different superstructures observed for oxygen order in the CuO chains of $YBa_2Cu_3O_{6+\delta}$

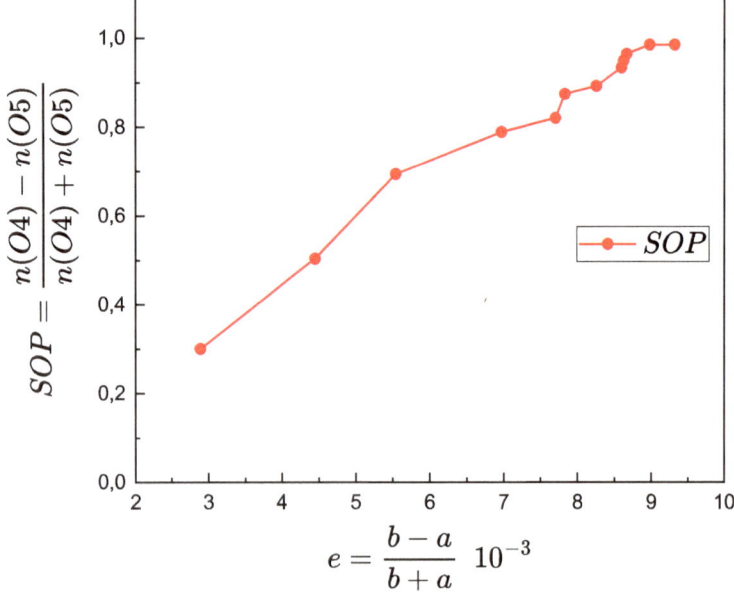

Fig. 2.14 SOP as a function of ϵ in the compound $YBa_2Cu_3O_{6+\delta}$ [50]

2.2.3 Irreversibility Line (LI)

Müller et al. [52] observed the irreversibility temperature T^* by measuring the magnetization M as a function of temperature in LaBaCuO. They obtained two curves for $T < T^*$: one during cooling in the field (F.C.: Field Cooled) and the other in the absence of the field (Z.F.C.: Zero Field Cooled). They coincide for $T \geq T^*$ (Fig. 2.15).

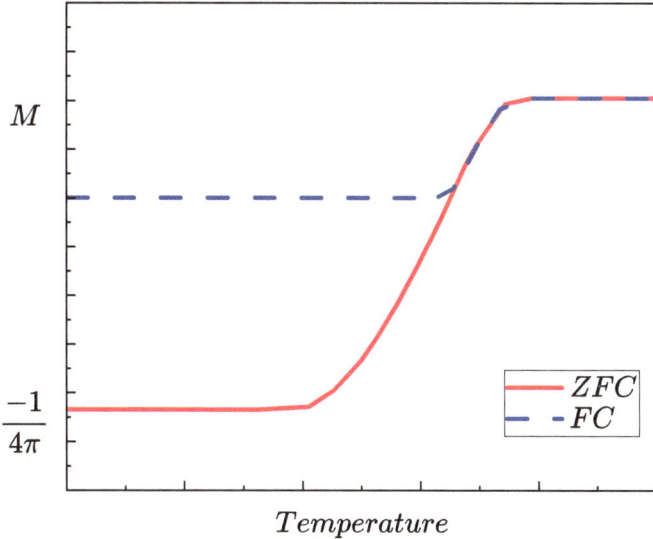

Fig. 2.15 Magnetization $M(T)$ showing the curves (Z.F.C.) and (F.C.)

Other measurements as a function of the field H give the curve $H^*(T)$ called the irreversibility line (I.L.). T^* ($T^* = T_{irr}$) decreases as H increases following the relation:

$$T^* = T_{irr}(H) = T_{irr}(0) \left(1 - \left(\frac{H}{K} \right)^{\frac{1}{n}} \right) \tag{2.1}$$

The coefficients K and n depend on the chosen model. For $YBa_2Cu_3O_{6+\delta}$, n can vary between 1.2 (thin films) and 1.7 (metal) [53]. $K = H(T^* = 0)$ is the magnetic field that cancels the inter-grain critical current at the limit T^* tends to 0. Thus, K sets the position of the I.L. in the H(T) plane [54].

The H(T) curve shows the irreversibility line of the polycrystal $YBa_2Cu_3O_{6+\delta}$ ($\delta \approx 1$) in Fig. 2.16. In the irreversible region, the sample exhibits irreversible magnetic behavior (in the sense of the hysteresis cycle) at low T and H. In the reversible region (reversible at high T and H), it exhibits reversible magnetic behavior (the critical current density is zero since the magnetization is perfectly reversible). The reversible region increases as the frequency increases because the pinning potential decreases with increasing field. The I.L. shifts toward lower temperatures as the amplitude of the magnetic field H_{ac} increases [55], and it delineates the "vortex glass" zone (zero resistivity) from the "liquid vortex" zone (non-zero resistivity). Between the irreversibility line and H_{c2}, the vortices are not anchored and will move within the sample. They form a liquid, hence the name "vortex liquid". However, below the irreversibility line, the vortices are anchored to the sample. The most interesting value is not H_{c2}, but H^*(irreversibility field), which is significantly lower.

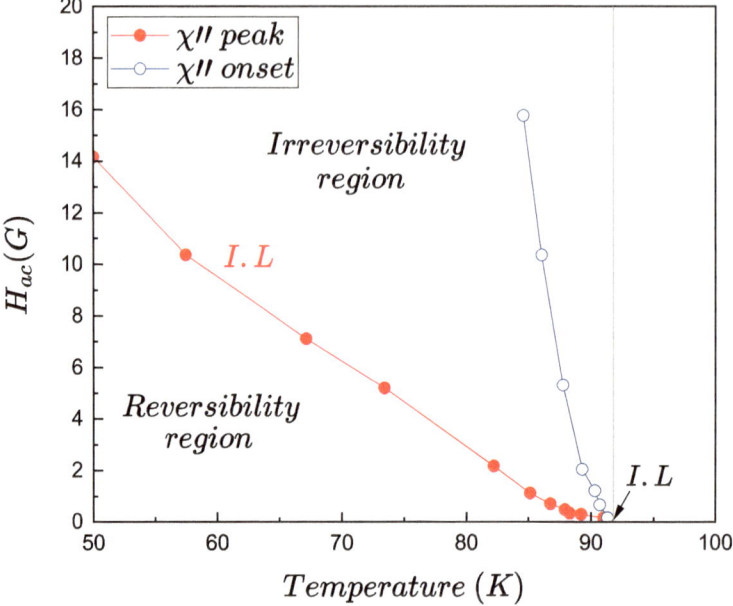

Fig. 2.16 Irreversibility line of the polycrystal YBa$_2$Cu$_3$O$_{6+\delta}$ [55]

Table 2.1 Effect of oxygen content δ on T_c, n, and k in YBa$_2$Cu$_3$O$_{6+\delta}$ [54]

z	T_c (K)	n (T)	K (KG)
1	89	1.5	342
0.85	75	1.5	101
0.7	58	1.5	48
0.6	39	1.6	23

On the irreversibility line, the critical current $J_c(T = T^* = T_{irr} = T_p) = 0$. In practice, the goal is to increase the irreversibility zone between the I.L. and H$_{c1}$(T). Our original contribution from the argon–oxygen thermal treatment, which increased K and T_c, of samples was studied in this direction.

Figure 2.17 illustrates the relationship between the AC magnetic field H_{ac} and the temperature T_p of the χ'' peak in Y123, determined for a frequency f of 500 Hz. For fields lower than 0.1 G (1 G $= 10^{-4}$ T), T_p is almost independent of the field and is close to the transition temperature $T_c = 90$ K. For fields higher than 1 G, T_p drops toward lower temperatures.

Figure 2.18 displays an increase in the irreversibility line with the partial oxygen content δ. For $\delta = 0.6$ ($T_c = 39$ K), the reversible region is very broad and begins to narrow as δ and T_c increase. Table 2.1 shows the effect of δ on the critical temperature T_c, the coefficient k, and the exponent n. The coefficient k decreases with δ, and the exponent n remains almost constant (Table 2.1).

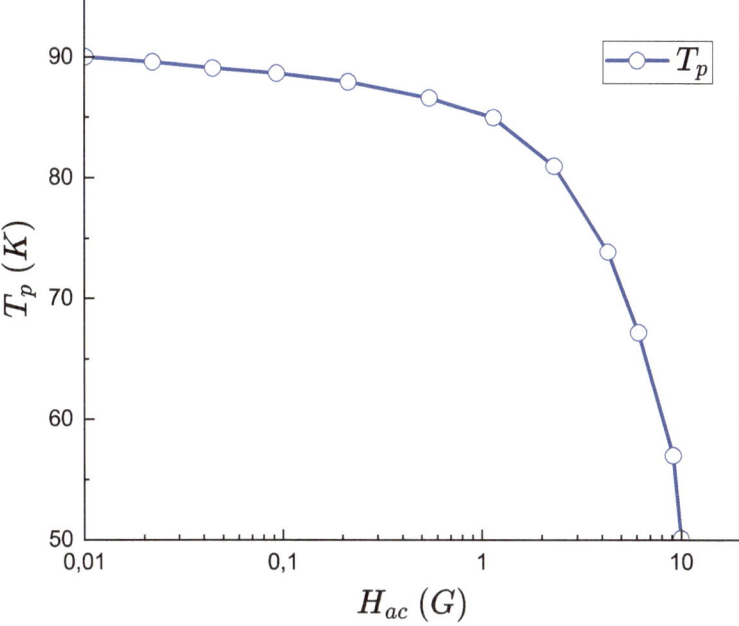

Fig. 2.17 The peak temperature T_p in YBa$_2$Cu$_3$O$_7$ as a function of the magnetic field H_{ac} [55]

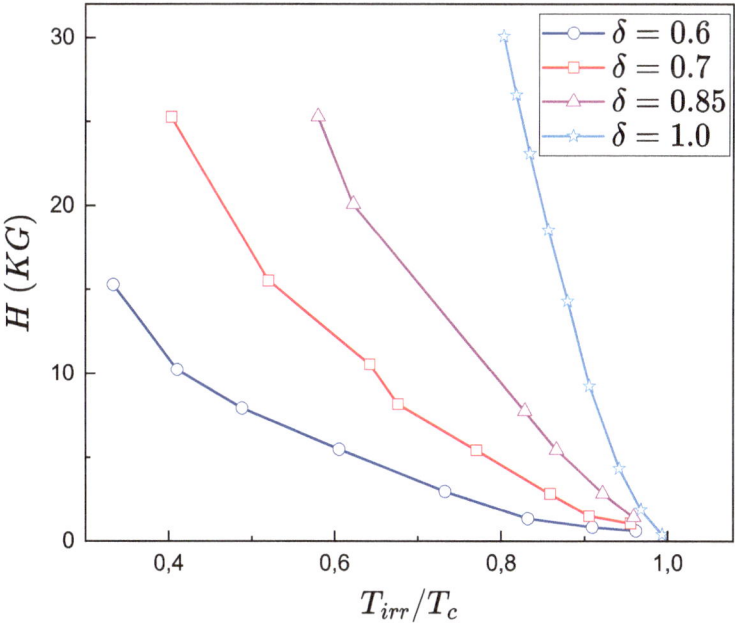

Fig. 2.18 Magnetic field as a function of $t = T_{irr}(H)/T_c(0)$ and the oxygen content z in YBa$_2$Cu$_3$O$_{6+\delta}$ [54]

2.2.4   Substitution in $YBaCu_3O_{6+\delta}$

As observed, substituting atoms in a crystal with foreign atoms is a widely used tool in solid-state physics to probe the properties of a superconductor. This is especially true in the case of YBaCuO, justified by the significant number and variety of publications produced on this material to date. Substitution in the YBaCuO compound can be carried out on the various cation sites (Y, Ba, Cu) [56], but only copper site substitution significantly affects superconducting properties since the CuO_2 planes are the locus of superconductivity.

2.2.4.1   Substitution at Copper Sites

Substitution on Copper Sites of the Plane: It's established that substitution at the Cu site is only possible if the replacing atoms are of similar size to copper, thereby capable of forming pyramidal, octahedral, or square planar coordinations with oxygen [57–59]. Nickel (Ni) and zinc (Zn) replace copper in the CuO_2 planes without inducing significant structural change and strongly reduce T_c. Figure 2.19, [60] shows a linear decrease of T_c, measured by alternative magnetic susceptibility $\chi_a c$, as x increases. For a given M_x, T_c decreases with δ as expected [61]. Moreover, comparison between doping with Ni and Zn shows that:

- T_c decreases more rapidly with the increase of Zn content than with Ni.
- Zn more rapidly destroys the superconducting phase, whereas the normal metallic phase is destroyed more quickly by Ni.

The substitution with magnetic Ni or non-magnetic Zn has a similar effect on superconductivity in cuprates. It is now well established that substituting Cu with Zn in the CuO_2 planes produces dramatic effects, such as a more significant decrease in T_c than any other substituting element for Cu [62], regardless of the zinc concentration, the structure remains orthorhombic, and the oxygen content stays constant [63, 64] or decreases slightly [65, 66].

Substitution on Copper Sites of the $Cu_{(1)}$-O Chains: Several ions can substitute for copper in superconducting oxides. It is accepted that, in $YBa_2Cu_3O_{6+\delta}$, trivalent ions Fe, Co, Ga, Al substitute preferentially at the Cu(1) site of the chains. This substitution influences the oxygen order in the basal plane, inducing an orthorhombic–tetragonal structural transition. Figure 2.20 shows a linear drop in T_c as x increases, with superconductivity disappearing for $x \geq 0.4$.

2.2.4.2   Substitution at Yttrium and Barium Sites

The main interest in adding another element to the YBCO compound can be summarized in two points: to vary its properties to find more information on possible superconductivity mechanisms and to improve physical characteristics such as density,

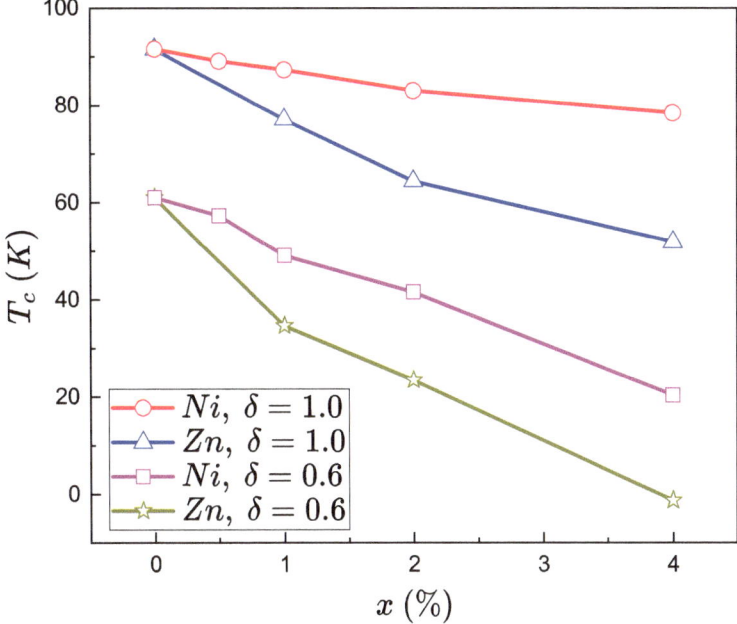

Fig. 2.19 T_c as a function of x in $YBa_2(Cu_{1-x}M_x)_3O_{6+\delta}$ [60]

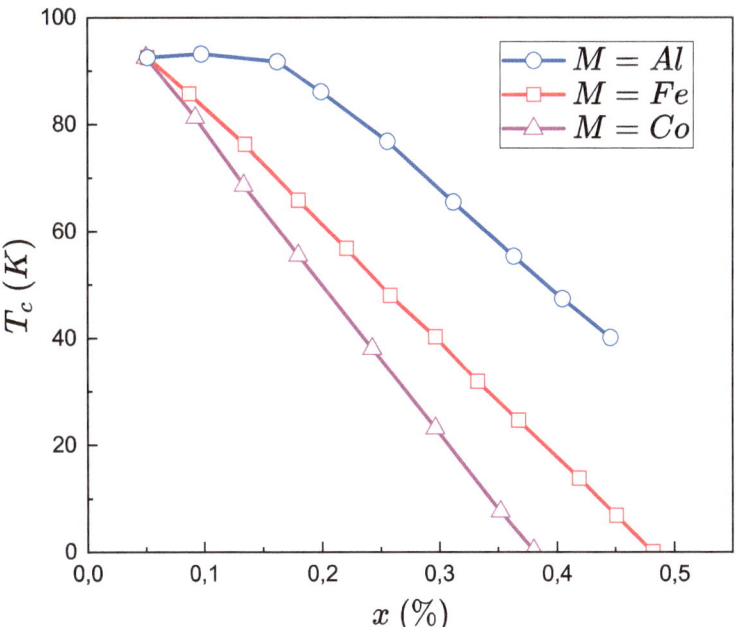

Fig. 2.20 T_c as a function of x in the compound $YBa_2(Cu_{3-x}M_x)O_{6+\delta}$ [67]

texture, or ductility. Substitution of Yttrium (Y) by rare earth elements [68–73] does not really affect the critical temperature, their crystal structure remains unchanged [74]. It has been noted that Yttrium sites are isolated from the superconducting region in this material. Conversely, substituting Lanthanum (La) in the Barium (Ba) site leads to a very rapid decrease in T_c. For example, superconductivity disappears in the compound $YBa_{2-x}La_xCu_3O_7$ for $x \geq 0.04$ [75, 76]. This effect is attributed either to charge compensation by the electron given to Lanthanum (by reducing the valence of Cu) or by increasing the oxygen content [77]. Further studies on substitution at the Y and Ba sites by rare earths Nd, Sm, Eu, Gd, and Dy indicate a significant decrease in T_c in the compound-type $RBa_{2-x}R_xCu_3O_7$, with total disappearance of superconductivity for $x = 0.05$ [77].

2.2.4.3 Study of the LnBaSrCu₃O₆₊δ System (Ln = rare earth)

In the compound $LnBa_2Cu_3O_{6+\delta}$, the transition temperature is not affected by the ionic radius of the rare earth [77]. This section discusses the structural and superconducting properties of the compound $LnBaSrCu_3O_{6+\delta}$. Studies, such as those by Wang et al. (with $z \approx 0.93$ and $z \approx 0.94$) [78], show that the critical temperature T_c varies with the ionic radius in the orthorhombic structure zone $r(\text{Ln}) < r(\text{Dy})$ and decreases as the ionic radius increases in the quadratic zone $r(\text{Ln}) > r(\text{Dy})$ (Fig. 2.21). However, for $r(\text{Ln}) = r(\text{Dy})$, two different structures and T_c are obtained, one orthorhombic with $T_c = 83$ K and the other quadratic with $T_c = 81$ K. This variation in structural and superconducting properties for the same compound is influenced by the heat treatment [79]. The highest critical temperature is observed for $GdBaSrCu_3O_{6+\delta}$ with $T_c = 86$ K [80].

Badri et al. (with $\delta = 0.9$) [81] found that a single phase is achieved from Ln = La to Ho. The Ln = Er case showed traces of impurities, while Ln = Tm, Yb, and Lu cases were multiphase. The crystal structure is quadratic for Ln = La, Gd, and orthorhombic for Ln = Dy and Ho.

Awana et al. [82] highlighted that T_c depends on the ionic radius in LnBaSrCu₃O₆₊δ (Ln = Y, Dy, Nd, and La). The structure is orthorhombic for Y and Dy with respective T_c of 81 K and 79 K. It is quadratic for Nd and La with T_c of 64 K and 45 K. The oxygen content determined from neutron diffraction spectra refinement was $\delta = 1.06$ for Y, 0.96 for Dy, 0.97 for La, and 1.01 for Nd.

The critical temperature T_c of $LnBaSrCu_3O_{6+\delta}$ (Ln = Gd, Dy, Eu) as a function of oxygen content is determined from electrical resistivity and magnetic susceptibility [83–85]. This figure shows that T_c strongly depends on the oxygen rate, with the highest temperatures observed for:

- $GdBaSrCu_3O_{6+\delta}$, $T_c = 86.2$ K for $\delta = 0.93$ [83] and 82 K for $\delta = 0.95$ [85].
- $DyBaSrCu_3O_{6+\delta}$, $T_c = 73.5$ K for $\delta = 0.9$ [84] and 81.4 K for $\delta = 0.97$ [85].
- $EuBaSrCu_3O_{6+\delta}$, $T_c = 72.9$ K for $\delta = 0.9$ [85].

The reasons for the variations in structural and superconducting properties when substituting Y with a rare earth Ln are not fully understood. In addition to the effect

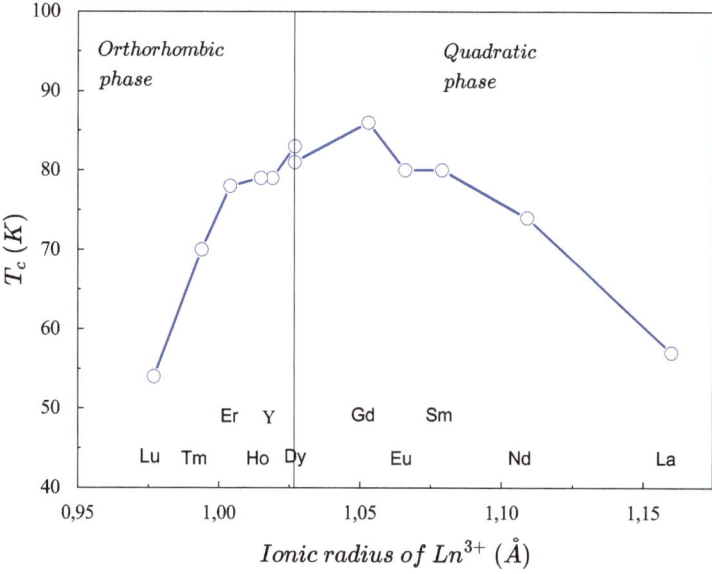

Fig. 2.21 Structural and superconducting properties in LnBaSrCu$_3$O$_{6+\delta}$ (Ln is a rare earth element) [78]

Table 2.2 Contradictory results for T_c obtained by various authors in the compound LnSr-BaCu$_3$O$_{6+\delta}$ (Ln = Eu, Sm, and Nd) [78, 81]

Ln	a (Å)	c (Å)	T_c (K)
Eu	3.844	11.579	80
Eu	3.845	11.59	60
Sm	3.859	11.60	37
Sm	3.851	11.591	80
Nd	3.87	11.622	74
Nd	3.865	11.64	58

of the ionic radius of the rare earth [80–82], it could be due to several factors such as the hole density, the degree of oxygen atom disorder, and the distance between the Cu(1) chain copper and the apical oxygen [83].

For LnSrBaCu$_3$O$_{6+\delta}$ ($\delta \approx 0.9$, Ln = Eu, Sm, Nd), Wang et al. [78] and Badri et al. [81] obtained quadratic structures and contradictory results for T_c in Table 2.2. To address these contradictions, we have explored the systems Y$_{0.5}$Ln$_{0.5}$BaSrCu$_3$O$_{6+\delta}$ with Ln = Eu, Sm, and Nd through X-ray diffraction, magnetic susceptibility, electrical resistivity, and iodometry. Additionally, we evaluated the significant effect of heat treatment with argon followed by annealing in oxygen [AO], which led to an increase in T_c, the irreversibility line, and magnetic shielding.

2.3   Conclusion

The exploration of high-temperature superconductivity in cuprate compounds, particularly those structured around the perovskite architecture, reveals a complex interplay of crystallography, chemistry, and physics. The critical temperature (T_c) at which these materials exhibit superconductivity is influenced by various factors, including the ionic radius of the constituent elements, oxygen content, and the presence of planes rich in copper and oxygen (CuO_2 planes), which are pivotal for the emergence of superconductivity. Substitutions within these compounds, whether at the copper sites or by replacing yttrium with rare earth elements, have profound effects on their superconducting properties, demonstrating the delicate balance required to optimize T_c.

Moreover, the introduction of elements such as lanthanum (La) and strontium (Sr) in place of barium (Ba) or within the lattice itself can significantly impact the material's superconducting phase, highlighting the role of electronic charge carriers and structural dimensionality in facilitating superconductivity. The study of these materials, through techniques like X-ray diffraction, magnetic susceptibility, and electrical resistivity, not only advances our understanding of superconductivity but also guides the synthesis of new compounds with potentially higher T_c and better performance characteristics.

Furthermore, the phenomena of irreversibility lines and the influence of thermal treatments on the oxygen content underscore the importance of microstructural control and the dynamic nature of these materials' superconducting states. This intricate relationship between structure, composition, and superconducting properties underscores the potential for discovering new high-T_c materials and the need for continued investigation into the mechanisms of superconductivity in cuprates and beyond.

As research progresses, the goal remains to unravel the complexities of high-temperature superconductivity, aiming for materials that operate at or near room temperature, which would revolutionize technology and energy systems. The insights gained from studying cuprates lay the groundwork for future advancements, promising a new era of superconducting applications that leverage the unique properties of these remarkable materials.

References

1. R.M. Hazen, L.W. Finger, R.J. Angel, C.T. Prewitt, N.L. Ross, H.K. Mao, C.G. Hadidiacos, P.H. Hor, R.L. Meng, C.W. Chu, Crystallographic description of phases in the Y-Ba-Cu-O superconductor. Phys. Rev. B **35**(13), 7238–7241 (1987)
2. G. Campi, A. Bianconi, Superstripes landscape in perovskites high T_c superconductors, in *Encyclopedia of Condensed Matter Physics*, 2nd edn., ed. by T. Chakraborty (Academic, Oxford, 2024), pp.437–447
3. M.H.K. Rubel, S.K. Mitro, B.K. Mondal, M.M. Rahaman, Md Saiduzzaman, J. Hossain, A.K.M.A. Islam, N. Kumada, Newly synthesized A-site ordered cubic-perovskite superconductor $(Ba_{0.54}K_{0.46})Bi_4O_{12}$: a dft investigation. Physica C: Supercond. Appl. **574**, 1353669 (2020)

4. Y. Cui, X. Chen, C. Li, L. Liu, D. Zhang, J. Li, Y. Tang, H. Tian, Theoretical investigation on thermodynamics and stability of anti-perovskite MgCNi$_3$ superconductor. Chem. Phys. Lett. **780**, 138961 (2021)
5. Y. Cai, W. Xie, H. Ding, Y. Chen, K. Thirumal, L.H. Wong, N. Mathews, S.G. Mhaisalkar, M. Sherburne, M. Asta, Computational study of halide perovskite-derived a2bx6 inorganic compounds: chemical trends in electronic structure and structural stability. Chem. Mater. **29**(18), 7740–7749 (2017)
6. M. Cyrot, D. Pavuna, *Introduction to Superconductivity and High-T$_c$ Materials* (World Scientific, 1992)
7. H. Rietschel, J. Fink, E. Gering, F. Gompf, N. Nocker, L. Pintschovius, B. Renker, W. Reichardt, H. Schmidt, W. Weber, Electronic and phononic properties of high-T$_c$ superconductors. Phys. C: Supercond. **153–155**, 1067–1071 (1988)
8. J. Bobroff, Etude par rmn des correlations magnetiques dans les supraconducteurs a haute temperature critique: effets des impuretes. Ph.D. thesis, PHDTHESIS, 1997. Thèse de doctorat dirigée par Alloul, Henri Physique Paris (1997)
9. I. Watanabe, N. Oki, T. Adachi, H. Mikuni, Y. Koike, F.L. Pratt, K. Nagamine, Muon spin relaxation study of the anomalous magnetic behavior in excess-oxygen-doped La$_{1.8}$Nd$_{0.2}$CuO$_{4+\delta}$. Phys. Rev. B **73**, 134506 (2006)
10. Y. Takeda, A. Sato, K. Yoshikawa, N. Imanishi, O. Yamamoto, M. Takano, Z. Hiroi, Y. Bando, Superconductivity of La$_{2-x}$A$_x$CuO$_{4+\delta}$ (A=Nd and Bi) prepared under high oxygen pressure. Phys. C: Supercond. **185–189**, 603–604 (1991)
11. A.R. Moodenbaugh, D.A. Fischer, Y.L. Wang, Y. Fukumoto, Superconductivity, oxygen content, and hole state density in Bi$_2$Sr$_{1.75}$Ca$_{1.25}$Cu$_2$O$_{8.18+y}$ ($-0.09 \leq y \leq$) and Bi$_{1.6}$Pb$_{0.4}$Sr$_{1.9}$Ca$_2$Cu$_3$O$_z$. Phys. C: Supercond. **268**(1), 107–114 (1996)
12. D.M. Pooke, G.V.M. Williams, Oxygen loading in (Bi, Pb) -2212 and -2223 materials. Phys. C: Supercond. **354**(1–4), 396–400 (2001)
13. C.K. Subramaniam, H.J. Trodahl, D. Pooke, K. Kishio, Thermoelectric power and the hole concentration in iodine-intercalated Bi$_2$Sr$_2$CaCu$_2$O$_x$ crystals. Phys. C: Supercond. **249**(1–2), 139–143 (1995)
14. D.G. Xenikos, P. Strobel, Critical-temperature optimization in Bi$_2$Sr$_2$CaCu$_2$O$_{8+\delta}$ crystals. Phys. C: Supercond. **248**(3–4), 343–348 (1995)
15. C. Michel, M. Hervieu, M.M. Borel, A. Grandin, F. Deslandes, J. Provost, B. Raveau, Superconductivity in the Bi-Sr-Cu-O system. Zeitschrift for Physik B Condensed Matter **68**(4), 421–423 (1987)
16. H. Maeda, Y. Tanaka, M. Fukutomi, T. Asano, A new high-T$_c$ oxide superconductor without a rare earth element. Jpn. J. Appl. Phys. **27**(2A), L209 (1988)
17. J.M. Tarascon, W.R. McKinnon, P. Barboux, D.M. Hwang, B.G. Bagley, L.H. Greene, G.W. Hull, Y. LePage, N. Stoffel, M. Giroud, Preparation, structure, and properties of the superconducting compound series Bi$_2$Sr$_2$Ca$_{n-1}$Cu$_n$O$_y$ with n = 1, 2, and 3. Phys. Rev. B **38**(13), 8885–8892 (1988)
18. K. Imai, I. Nakai, T. Kawashima, S. Sueno, A. Ono, Single crystal X-Ray structure analysis of Bi$_2$(Sr, Ca)$_2$CuO$_x$ and Bi$_2$(Sr, Ca)$_3$Cu$_2$O$_x$ superconductors. J. J. Appl. Phys. **27**(9A), L1661 (1988)
19. E. Takayama-Muromachi, Y. Uchida, Y. Matsui, M. Onoda, K. Kato, On the 110 K superconductor in the Bi-Ca-Sr-Cu-O system. Jpn. J. Appl. Phys. **27**(4A), L556 (1988)
20. A.C. Mark, M. Ahart, R. Kumar, C. Park, Y. Meng, D. Popov, L. Deng, C.-W. Chu, J.C. Campuzano, R.J. Hemley, Structure and equation of state of Bi$_2$Sr$_2$Ca$_{n-1}$Cu$_n$O$_{2n+4+\delta}$ from X-Ray diffraction to megabar pressures. Phys. Rev. Mater. **7**(6) (2023)
21. Z.Z. Sheng, A.M. Hermann, Superconductivity in the rare-earth-free Tl-Ba-Cu-O system above liquid-nitrogen temperature. Nature **332**(6159), 55–58 (1988)
22. Z.Z. Sheng, A.M. Hermann, Bulk superconductivity at 120 K in the Tl-Ca/Ba-Cu-O system. Nature **332**(6160), 138–139 (1988)
23. R.M. Hazen, L.W. Finger, R.J. Angel, C.T. Prewitt, N.L. Ross, C.G. Hadidiacos, P.J. Heaney, D.R. Veblen, Z.Z. Sheng, A. El Ali, A.M. Hermann, 100-k superconducting phases in the Tl-Ca-Ba-Cu-O system. Phys. Rev. Lett. **60**(16), 1657–1660 (1988)

24. S.S.P. Parkin, V.Y. Lee, E.M. Engler, A.I. Nazzal, T.C. Huang, G. Gorman, R. Savoy, R. Beyers, Bulk superconductivity at 125 K in $Tl_2Ca_2Ba_2Cu_3O_x$. Phys. Rev. Lett. **60**(24), 2539–2542 (1988)
25. E. Bellingeri, R. Flükiger, *TIBCCO* (CRC Press, 2022), pp. 260–278
26. M. Paranthaman, A.M. Hermann, *Related Topics II: Thallium-Based High-T$_c$ Superconducting Oxides* (CRC Press, 2022), pp. 735–757
27. R. Castillo, J. Cisterna, I. Brito, S. Conejeros, J. Llanos, Structure and properties of $(CH_3NH_3)_3Tl_{l2}Cl_9$: a thallium-based hybrid perovskite-like compound. Inorg. Chem. **59**(14), 9471–9475 (2020)
28. Y. Shimakawa, Y. Kubo, T. Manako, Y. Nakabayashi, H. Igarashi, Rietveld analysis of $Tl_2 Ba_2 Ca_{n-1} Cu_n O_{4+2n}$ (n=1, 2 and 3) by powder X-Ray diffraction. Phys. C: Supercond. **156**(1), 97–102 (1988)
29. D.K. Namburi, D.A. Cardwell, *Design of Cuprate HTS Superconductors* (Springer Nature Singapore, Singapore, 2022), pp. 239–270
30. R.G. Sharma, *Practical Cuprate Superconductors* (Springer International Publishing, Cham, 2021), pp. 227–275
31. S.N. Putilin, I. Bryntse, E.V. Antipov, New complex copper oxides: $HgBa_2RCu_2O_7$ (R = La, Nd, Eu, Gd, Dy, Y). Mate. Res. Bull. **26**(12), 1299–1307 (1991)
32. S.S.P. Parkin, V.Y. Lee, A. Nazzal, R. Savoy, T.C. Huang, G. Gorman, R. Beyers, Model family of high-temperature superconductors: $Tl_mCa_{n-1} Ba_2 Cu_n O_{2(n+1)+m}$ (m = 1, 2; n = 1, 2, 3). Phys. Rev. B **38**(10), 6531–6537 (1988)
33. S.N. Putilin, E.V. Antipov, O. Chmaissem, M. Marezio, Superconductivity at 94 K in $HgBa_2CuO_{4+\delta}$. Nature **362**(6417), 226–228 (1993)
34. S.N. Putilin, E.V. Antipov, M. Marezio, Superconductivity above 120 K in $HgBa_2CaCu_2O_{6+\delta}$. Phys. C: Supercond. **212**(3–4), 266–270 (1993)
35. L. Wang, X. Luo, J. Li, J. Zeng, M. Cheng, J. Freyermuth, Y. Tang, B. Yu, G. Yu, M. Greven, Y. Li, Growth and characterization of $HgBa_2CaCu_2O_{6+\delta}$ and $HgBa_2Ca_2Cu_3O_{8+\delta}$ crystals. Phys. Rev. Mater. (2018)
36. A. Schilling, M. Cantoni, J.D. Guo, H.R. Ott, Superconductivity above 130 K in the Hg-Ba-Ca-Cu-O system. Nature **363**(6424), 56–58 (1993)
37. M.K. Wu, J.R. Ashburn, C.J. Torng, P.H. Hor, R.L. Meng, L. Gao, Z.J. Huang, Y.Q. Wang, C.W. Chu, Superconductivity at 93 K in a new mixed-phase Y-Ba-Cu-O compound system at ambient pressure. Phys. Rev. Lett. **58**(9), 908–910 (1987)
38. A.C. Mark, J.C. Campuzano, R.J. Hemley, Progress and prospects for cuprate high temperature superconductors under pressure. High Press. Res. **42**(2), 137–199 (2022)
39. J.M. Tranquada, A.H. Moudden, A.I. Goldman, P. Zolliker, D.E. Cox, G. Shirane, S.K. Sinha, D. Vaknin, D.C. Johnston, M.S. Alvarez, A.J. Jacobson, J.T. Lewandowski, J.M. Newsam, Antiferromagnetism in $YBa_2 Cu_3 O_{6+x}$. Phys. Rev. B **38**(4), 2477–2485 (1988)
40. H. Alloul, P. Mendels, H. Casalta, J.F. Marucco, J. Arabski, Correlations between magnetic and superconducting properties of Zn-substituted $YBa_2Cu_3O_{6+x}$. Phys. Rev. Lett. **67**(22), 3140–3143 (1991)
41. J. Bobroff, Etude par rmn des correlations magnetiques dans les supraconducteurs a haute temperature critique: effets des impuretes. Ph.D. thesis, PHDTHESIS, 1997. Thèse de doctorat dirigée par Alloul, Henri Physique Paris (1997)
42. F. Jin, H. Zhang, W. Wang, X. Liu, Q. Chen, Improvement in structure and superconductivity of $YBa_2Cu_3O_{6+\delta}$ ceramics superconductors by optimizing sintering processing. J. Rare Earths **35**(1), 85–89 (2017)
43. C. De Boeck, Thèse de doctorat en science. Ph.D. dissertation, Université libre de Bruxelles (1997). [II. 29]
44. M. Milic, V. Matic, N. Lazarov, The dependence of critical temperature on oxygen concentration in $YBa_2Cu_3O_{6+x}$ in terms of the fragmented chain model. Open Phys. **9**(3) (2011)

45. Min Zhou, Aimin Sun, Lina Sun, Peipei Han, Effect of the oxygen distribution and inhomogeneity on the superconducting properties of $YBa_2Cu_3O_{7-\delta}$ $+xmol$ % $Y_2Ba_4CuMoO_y$ superconductors. Supercond. Sci. Technol. **26**(10), 105006 (2013)
46. H.J. Zhang, X.P. Zhang, J.P. Shi, H.F. Tian, Y.G. Zhao, Effect of oxygen content and superconductivity on the nonvolatile resistive switching in $YBa_2Cu_3O_{6+x}$ / Nb-doped $SrTiO_3$ heterojunctions. Appl. Phys. Lett. **94**(9) (2009)
47. M. Bushgeister, W. Hiller, S.M. Hosseini, K. Kopitzki, D. Wagener, Transport properties of superconductors: Proceedings of the ictps '90 international conference, in *Transport Properties of Superconductors* (World Scientific, 1990)
48. P. Raju, Gupta, Michèle Gupta, Relationship between radiation-induced orthorhombic-tetragonal phase transformation and loss of superconductivity in $YBa_2Cu_3O_7$. Phys. Rev. B **45**(17), 9958–9965 (1992)
49. H. Luetgemeier, I. Heinmaa, D. Wagener, S.M. Hosseini, *Superconductivity Versus Oxygen Concentration in 123 Compounds: Influence of RE Ionic Radii Studied by Cu NQR* (Springer, Berlin Heidelberg, 1994), pp. 225–235
50. J.D. Jorgensen, D.G. Hinks, P.G. Radaelli, S. Pei, P. Lightfoot, B. Dabrowski, C.U. Segre, B.A. Hunter, Defects, defect ordering, structural coherence and superconductivity in the 123 copper oxides. Phys. C: Supercond. 185–189:184–189 (1991)
51. H. Shaked, J.D. Jorgensen, J. Faber, D.G. Hinks, B. Dabrowski, Theory for oxygen content and ordering in $YBa_2Cu_3O_{6+x}$ in equilibrium with oxygen gas. Phys. Rev. B **39**(10), 7363–7366 (1989)
52. K.A. Müller, M. Takashige, J.G. Bednorz, Flux trapping and superconductive glass state in $La_2CuO_{4-y}Ba$. Phys. Rev. Lett. **58**(11), 1143–1146 (1987)
53. Y. Xu, M. Suenaga, Irreversibility temperatures in superconducting oxides: the flux-line-lattice melting, the glass-liquid transition, or the depinning temperatures. Phys. Rev. B **43**(7), 5516–5525 (1991)
54. J. Vanacken, E. Osquiguil, Y. Bruynseraede, Irreversibility line and critical currents in oxygen deficient $YBa_2Cu_3O_x$ ceramics. Phys. C: Supercond. **183**(1–3), 163–166 (1991)
55. N. Savvides, A. Katsaros, C. Andrikidis, K.-H. Müller, Temperature, field and frequency dependence of intergranular ac loss in high-temperature superconductors. Phys. C: Supercond. **197**(3–4), 267–273 (1992)
56. J.M.S. Skakle, Crystal chemical substitutions and doping of $YBa_2Cu_3O_x$ and related superconductors. Mater. Sci. Eng.: R: Rep. **23**(1), 1–40 (1998)
57. B. Raveau, C. Michel, M. Hervieu, D. Groult, *Crystal Chemistry of High-T_c Superconducting Copper Oxides* (Springer, Berlin Heidelberg, 1991)
58. A.V. Egorov, E. Selimović, A.V. Komolkin, T. Soldatović, Substitution behavior of square-planar and square-pyramidal Cu(II) complexes with bio-relevant nucleophiles. J. Coord. Chem. **71**(7), 1003–1019 (2018)
59. T. Ateş Turkmen, L. Zeng, Y. Cui, İ. Fidan, F. Dumoulin, C. Hirel, Y. Zorlu, V. Ahsen, A.A. Chernonosov, Y. Chumakov, K.M. Kadish, A.G. Gürek, S. Tokdemir Öztürk, Effect of the substitution pattern (peripheral vs non-peripheral) on the spectroscopic, electrochemical, and magnetic properties of octahexylsulfanyl copper phthalocyanines. Inorg. Chem. **57**(11), 6456–6465 (2018)
60. J. Bobroff, Etude par rmn des correlations magnetiques dans les supraconducteurs a haute temperature critique: effets des impuretes. Ph.D. thesis, Paris 11, 1997. Thèse de doctorat dirigée par Alloul, Henri Physique Paris 11 (1997)
61. P. Mendels, H. Alloul, J.H. Brewer, G.D. Morris, T.L. Duty, S. Johnston, E.J. Ansaldo, G. Collin, J.F. Marucco, C. Niedermayer, D.R. Noakes, C.E. Stronach, Muon-spin-rotation study of the effect of Zn substitution on magnetism in $YBa_2Cu_3O_x$. Phys. Rev. B **49**(14), 10035–10038 (1994)
62. M. Rubinstein, D.J. Gillespie, J.E. Snyder, T.M. Tritt, Effects of Gd, Co, and Ni doping in $La_{2/3}Ca_{1/3}MnO_3$ resistivity, thermopower, and paramagnetic resonance. Phys. Rev. B **56**(9), 5412–5423 (1997)

63. H. Zhang, X.Y. Zhao, Y. Zhao, S.H. Liu, Q.R. Zhang, Oxygen content is not the predominant factor for high T_c superconductivity in YBaCuO system. Solid State Commun. **72**(1), 75–79 (1989)
64. C. Lin, Z.-X. Liu, J. Lan, Effect of Ni and Zn substitution on magnetic properties of the high-T_c superconductor $GdBa_2Cu_3O_{7-y}$. Phys. Rev. B **42**(4), 2554–2557 (1990)
65. P.H. Andresen, H. Fjellvåg, P. Karen, A. Kjekshus, K.S. Varma, J. Becher, A.E. Underhill, Substitution for copper in $YBa_2Cu_3O_{9-\delta}$ by $3d^-$ and pre-transition metals. Acta Chemica Scandinavica **45**, 698–708 (1991)
66. R.S. Roth, C.J. Rawn, F. Beech, J.D Whitler, J.O. Anderson, Phase equilibria in the system Ba-Y-Cu-O-CO_2 in air, in *Ceramic Superconductors II* (Society of Petroleum Engineers, 1988)
67. J.M. Tarascon, P. Barboux, P.F. Miceli, L.H. Greene, G.W. Hull, M. Eibschutz, S.A. Sunshine, Structural and physical properties of the metal M substituted $YBa_2Cu_{3-x}M_xO_{7-y}$ perovskite. Phys. Rev. B **37**(13), 7458–7469 (1988)
68. J. Shimoyama, Y. Nakayama, K. Kitazawa, K. Kishio, Z. Hiroi, I. Chong, M. Takano, Strong flux pinning up to liquid nitrogen temperature discovered in heavily pb-doped and oxygen controlled Bi-2212 single crystals. Phys. C: Supercond. **281**(1), 69–75 (1997)
69. I. Chong, Z. Hiroi, M. Izumi, J. Shimoyama, Y. Nakayama, K. Kishio, T. Terashima, Y. Bando, M. Takano, High critical-current density in the heavily Pb-Doped $Bi_2Sr_2CaCu_2O_{8+\delta}$ superconductor: generation of efficient pinning centers. Science **276**(5313), 770–773 (1997)
70. D.T. Verebelyi, C.W. Schneider, Y.-K. Kuo, M.J. Skove, G.X. Tessema, J.E. Payne, Effect of magnetic substitutions (Ni Co, Fe) for Cu on thermal conductivity of BiSCCO whiskers. Phys. C: Supercond. **328**(1–2), 53–59 (1999)
71. F. Herman, R.V. Kasowski, W.Y. Hsu, Electronic structure of $Bi_2Sr_2CaCu_2O_8$ high-tcsuperconductors. Phys. Rev. B **38**(1), 204–207 (1988)
72. M.K. Yu, J.P. Franck, Specific heat of Zn and Co substituted $Bi_{1.8}Pb_{0.2}Sr_2Ca(Cu_{1-x}M_x)_2O_y$. Phys. Rev. B **48**(18), 13939–13944 (1993)
73. B.A. Richert, R.E. Allen, Atomic substitutions in $YBa_2Cu_3O_7$: modification of the electronic structure. Phys. Rev. B **37**(13), 7496–7501 (1988)
74. R. Kuentzler, S. Vilminot, Y. Dossmann, A. Derory, Superconductivity and low temperature specific heat in $YBa_2(Cu_{1-x}M_x)_3O_{7-\delta}$ systems (M =Zn and Fe). Phys. C: Supercond. **153–155**, 1032–1033 (1988)
75. J. Toulouse, R. Pattnaik, Relaxor and superparaelectric behavior in the disordered ferroelectrics KLT and KTN. Ferroelectrics **199**(1), 287–305 (1997)
76. D. Viehland, M. Wuttig, L.E. Cross, The glassy behavior of relaxor ferroelectrics. Ferroelectrics **120**(1), 71–77 (1991)
77. J.C.L. Chow, H.-T. Leung, W. Lo, D.A. Cardwell, Effects of pt doping on the size distribution and uniformity of particles in large-grain YBCO. Supercond. Sci. Technol. **11**(4), 369–374 (1998)
78. X.Z. Wang, B. Hellebrand, D. Bäuerle, Crystal structure and superconductivity in REBaSrCu_3O_x. Phys. C: Supercond. **200**(1–2), 12–16 (1992)
79. X.Z. Wang, P.L. Steger, M. Reissner, W. Steiner, Oxygen ordering and superconductivity in $DyBaSrCu_3O_y$. Phys. C: Supercond. **196**(3–4), 247–251 (1992)
80. X.Z. Wang, D. Bäuerle, High-T_c superconductivity in $GdBaSrCu_3O_7$. Phys. C: Supercond. **176**(4–6), 507–510 (1991)
81. S.K. Malik, S.S. Shah, *Physical and Material Properties of High Temperature Superconductors* (Nova Publishers, 1994)
82. V.P.S. Awana, C.A. Cardoso, O.F. de Lima, S.K. Malik, W.B. Yelon, Ram Prasad, A. Gupta, A. Sedky, A.V. Narlikar, Rare earth ionic size dependence of T_c in $RBaSrCu_3O_7$ (R = Y, Dy, Nd, and La) series. Phys. C: Supercond. 341–348:627–628 (2000)
83. V.P. Dyakonov, V.I. Markovich, R. Puzniak, H. Szymczak, N.A. Doroshenko, Y.I. Yuzhelevskii, Influence of strontium on superconducting and magnetic properties of $DyBa_{2-x}Sr_xCu_3O_{7-\delta}$. Phys. C: Supercond. **225**(1–2), 51–58 (1994)

84. V.P. Dyakonov, L.P. Kozeeva, G.G. Levchenko, V.I. Markovich, A.A. Pavljuk, J.F. Revenko, I.M. Fita, Low-temperature structural phase transition of the jahn-teller type in $DyBa_2Cu_3O_{7-\delta}$. Phys. C: Supercond. **208**(1–2), 23–28 (1993)
85. K. Yamaya, Y. Okajima, T. Yagi, M. Domon, F. Itoh, Superconductivity in the $YBa_2Cu_3O_y$-like compounds with the disordered BaO plane. Phys. C: Supercond. **185–189**, 1237–1238 (1991)

Chapter 3
Experimental Techniques for YBCO System Characterization

3.1 Introduction

The preparation of samples, particularly in the field of superconductivity, involves a variety of methods tailored to achieve specific electrical and magnetic applications. These methods range from polycrystalline powders, compacted ceramics, to thin or thick films, each aiming to control crystalline sizes, specific surfaces, and grain boundaries to optimize superconducting properties.

Among the fabrication techniques for superconducting materials, spin coating for centrifugation evaporation, sputtering for cathodic spraying, and laser ablation for thin film deposition are prominent. Sol–gel methods or spray pyrolysis are also employed for creating polycrystalline samples or thin films. The development of monocrystalline or bulk samples requires conventional metallurgy techniques (Table 3.1).

The solid-state reaction or solid-state synthesis is the most commonly used method for producing ceramic superconductors. This involves mixing oxides and carbonates in powder form and reacting them through thermal treatment after multiple grinding stages. The solid-state route is favored for its simplicity and quick implementation, as it doesn't require any precursor preparation. It has been particularly successful for certain superconductors like YBCO. However, this technique can encounter several issues [1].

The polycrystalline samples $Y_{0.5}Ln_{0.5}SrBaCu_3O_{6+\delta}$ (Ln = Eu, Sm, and Nd) were prepared by high-temperature solid-state reaction using highly pure CuO, Ln_2O_3 oxides, and barium and strontium carbonates ($BaCO_3$ and $SrCO_3$). The preparation process involved the following steps:

- Weighing the starting oxides and carbonates in appropriate amounts.
- Mixing the weighed powders.
- Grinding the mixture for varying durations.
- Calcination to eliminate carbon and achieve the desired phase with high purity.

K. Khallouq, *Exploring High-Temperature Superconductivity in the YBCO System*, SpringerBriefs in Materials, https://doi.org/10.1007/978-3-031-66238-6_3

Table 3.1 Table of possible issues and their causes in material preparation

Possible issues	Possible causes
Homogeneity defects	Poorly prepared mixture, particles too large, poor diffusion
Many parasitic phases (impurities)	Weighing inaccuracies, incomplete reaction (insufficient temperature maintenance or too low temperature)
Poor distribution of constituents	Poor mixture homogeneity, ineffective grinding

- Grinding followed by pressing into pellets under varying pressures.
- One or more sintering stages to increase the volumetric fraction of the desired phase.

3.2 Sample Preparation

3.2.1 Introduction

For applications requiring electrical and magnetic properties, numerous methods exist for the fabrication of functional polycrystalline oxides, whether as powders, compacted ceramics, or thin or thick films. The goal of each method is to control crystallite sizes, specific surface areas, and grain boundaries to optimize superconducting properties.

Techniques such as spin coating, sputtering, and laser ablation are utilized for creating thin superconducting films. Similarly, sol–gel and spray pyrolysis methods are employed for producing polycrystalline samples or thin films. The fabrication of monocrystalline or bulk samples demands conventional metallurgy techniques.

The solid-state reaction, or solid-state synthesis, is widely used for crafting ceramic superconductors. It involves mixing oxides and carbonates in powder form and subjecting them to thermal treatment after extensive grinding. This technique is favored for its simplicity and efficiency, particularly for certain superconductors like YBCO. However, this method may present challenges [1], as outlined in the following table:

Polycrystalline samples $Y_{0.5}Ln_{0.5}SrBaCu_3O_{6+\delta}$ (Ln = Eu, Sm, and Nd) were prepared via high-temperature solid-state reaction using pure oxides CuO, Ln_2O_3 and carbonates $BaCO_3$, $SrCO_3$. The fabrication followed these steps:

- Weighing the oxides and carbonates in precise amounts.
- Mixing the weighed powders.
- Grinding the mixture to enhance homogeneity.
- Calcination to remove carbon and synthesize the desired phase.

- Grinding followed by pressing into pellets under specific pressure.
- One or more sintering processes to increase the volumetric fraction of the desired phase.

3.2.2 Preparation of Mixtures

The initial step involves accurately weighing and mixing the starting materials: Y_2O_3, Ln_2O_3, $BaCO_3$, $SrCO_3$, and CuO, according to the stoichiometry of the general formula $Y_{0.5}Ln_{0.5}SrBaCu_3O_{6+\delta}$ (Ln = Eu, Sm, or Nd), that is, Y/Ln/Ba/Sr/Cu $= (1-x)/(x)/(1)/(1)/(3)$ as per the designated reaction.

3.2.3 Mixing and Grinding

Mixing and grinding are crucial for achieving a homogeneous mixture. The stoichiometric blend of Y_2O_3, Ln_2O_3, $BaCO_3$, $SrCO_3$, and CuO is manually ground in an agate mortar. This process aims to reduce particle size and enhance mixture uniformity.

3.2.4 Calcination

Calcination serves as a preparatory step for sample fabrication, involving the decomposition of carbonates by heating in air or a specific atmosphere below $1100\,°C$, specifically at $950\,°C$, the decomposition temperature of the material, for 18 h. This step leads to various reactions and changes in the sample.

3.2.5 Grinding and Compacting

Post-calcination, the resulting black oxide composite block is re-ground and compacted into pellets using a uniaxial press mold (Fig. 3.1) under a pressure of 10 Tons/cm². This compaction is carefully done in stages to prevent air entrapment.

Fig. 3.1 Diagram of a
uniaxial pressing mold

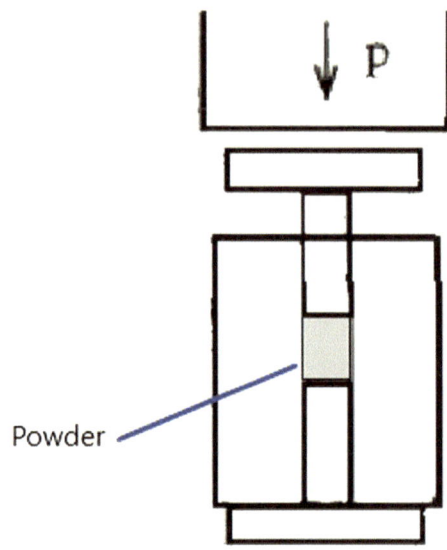

3.2.6 Sintering

Sintering, a thermal process conducted in air or controlled atmospheres, is performed at temperatures between 950 °C and 980 °C for 24 h. This process enables grain inter-diffusion and the formation of monocrystalline phase grains, which bond together to allow macroscopic current flow. Sintering consolidates the material, reduces porosity, and increases grain size.

After this second heating, a final cycle of grinding, compaction, and sintering ensures sample homogeneity and proper grain contact, yielding cylindrical pellets with dimensions of 1 mm thickness and 10 mm diameter.

3.2.7 Thermal Treatments

3.2.7.1 Oxygenation Treatment: [O] Sample

The final step involves annealing under an oxygen flow in a furnace at a fixed temperature of 450 °C for 72 h. This process increases the oxygen content of the sample to $6 + \delta \approx 7$.

3.2.7.2   Argon Treatment: [AO] Sample

Another thermal treatment was performed on the same [O] sample. Annealing under argon for 18 h at $T = 850\,°C$, followed by annealing under oxygen for 72 h at $T = 450\,°C$. The cooling occurs slowly in several steps down to room temperature.

The various steps followed for synthesis via the solid-state reaction are outlined in the flowchart in Fig. 3.2.

Various techniques can be used to characterize the samples from structural, electrical, and magnetic perspectives:

- X-ray diffraction (XRD).
- Iodometry measurements.
- Electrical resistivity measurements as a function of temperature $\rho(T)$.
- Magnetic susceptibility measurements.

3.3   X-Ray Diffraction

3.3.1   Equipment Used

X-ray diffraction is a characterization method that identifies the crystalline phases of a solid [2, 3]. It tracks the evolution of the unit cell, crystallinity, and size of crystallites as a function of treatments applied to solids. It's widely used to analyze thin film growth, phases present in powders or bulk materials, and the arrangement of atoms in crystal lattices. This method is only applicable to mediums exhibiting the characteristics of a crystalline state: a periodic and ordered arrangement.

X-rays, besides easily penetrating matter, interact with matter in a specific way, enabling chemical and structural analysis of a given material. Generally, the principle of analysis is as follows: the sample to be characterized is irradiated with an X-ray beam. The sample then diffracts waves in a spatial distribution of intensity, providing information on the material's structure.

X-ray diffraction involves bombarding the sample with X-rays (the sample must be a homogenous isotropic powder or a solid composed of small crystals fused together), then measuring the intensity of the X-rays scattered according to the crystallographic orientation of the crystal in space.

The recording of diffracted X-ray intensity spectra as a function of the incidence angle θ relative to the sample's surface is performed using an X-ray diffractometer. The diffracted radiation is measured by a sensor counter, and the sample is mounted on a goniometer, either a quarter-circle for crystals or simpler for powders and polycrystals [4] (Fig. 3.3). Information is derived from techniques and models initiated by Bragg, who considered a single crystal as a diffraction grid and a monochromatic X-ray beam to deduce his famous formula.

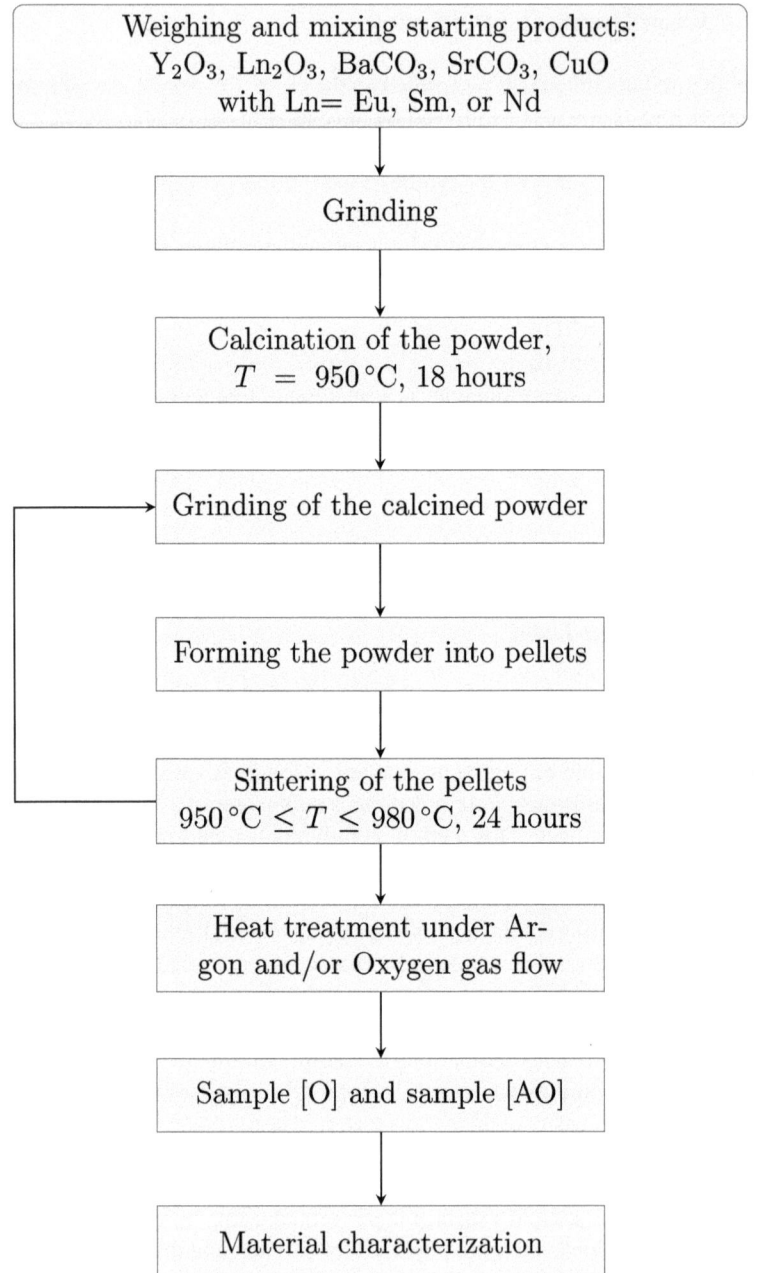

Fig. 3.2 Flowchart of the sample preparation process for $Y_{0.5}Ln_{0.5}SrBaCu_3O_{6+\delta}$

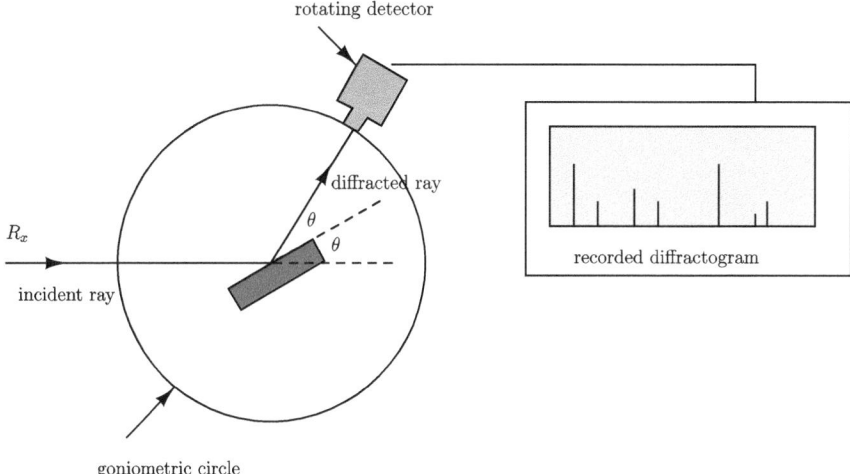

Fig. 3.3 Principle diagram of an X-ray diffractometer

Generally, this method is used to

- Determine the chemical composition of the powder by comparing the obtained spectrum with those in the PDF (powder diffraction file) database of the ICDD (International Centre for Diffraction Data).
- Study the crystal parameters a, b, c, α, β, γ.
- Determine the atomic positions and space group.
- Determine the interatomic distances.
- Detect the presence of impurities.

In our research, we used radiations with a wavelength $\lambda_{K\alpha} = 1.5418$ Å emitted by a copper $K\alpha$ anticathode, bombarded by electrons (emitted by a tungsten filament) accelerated under a voltage of 40 kV, I = 30 mA. The diffractometer is equipped with slits in front of the rotating sample holder and a monochromator. It is controlled by a computer through a data acquisition system for diffraction, and the results are presented in the form of a detailed structural map of the crystal's unit cell, showing the relative locations of all its atoms.

3.3.2 Bragg's Law

Bragg's law governs the diffraction of electromagnetic waves by a crystal (Fig. 3.4). It connects the distance between atoms in a crystal and the angles at which X-rays sent to the crystal are primarily diffracted. Considering an X-ray beam of wavelength λ incident at an angle θ on a family of crystallographic planes (*hkl*) defined by their interplanar distance d_{hkl} (i.e., the distance between two crystallographic planes),

Fig. 3.4 Diagram
illustrating X-ray diffraction
on lattice planes

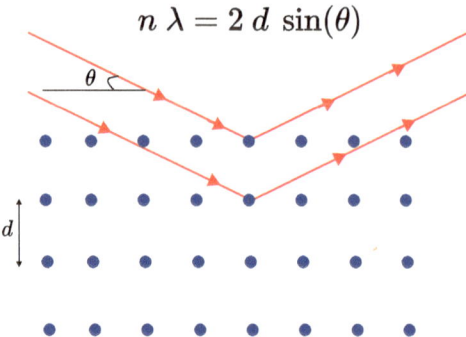

diffraction occurs when the path difference between the incident and diffracted rays
by the atoms equals a multiple of the wavelength.

Constructive interference occurs when Bragg's Law is satisfied [5]:

$$2.d_{hkl} sin(\theta_{hkl}) = n\lambda \tag{3.1}$$

where n is the diffraction order (integer) and

$$d_{hkl} = 2\frac{\pi}{OG} \text{ with } \vec{OG} = h\vec{a^*} + k\vec{b^*} + l\vec{c^*} \tag{3.2}$$

$\vec{a^*}, \vec{b^*}$, and $\vec{c^*}$ are the reciprocal lattice parameters. For an orthorhombic structure
($a \neq b \neq c$ and $\alpha = \beta = \gamma = 90\,°C$):

$$1/d_{hkl} = (h/a)^2 + (k/b)^2 + (l/c)^2 \tag{3.3}$$

Using the experimental Bragg angles θ_{hkl}, with Eqs. (3.1) and (3.3), we can deduce
the a, b, and c parameters of the crystal lattice.

3.3.3 Diffracted Intensity

Since the unit cell contains several atoms, to determine the diffracted amplitude in a
given direction, we must sum the complex amplitudes fa (atomic scattering factor)
scattered by the atoms. This diffracted amplitude by all the crystal's atoms is called
the structure factor and is written as

$$F_{hkl} = \sum_j N_j \, f_j \, exp(-B_j(\sin\theta/\lambda)^2) \, exp(2i\pi(hx_j + ky_j + lz_j)) \tag{3.4}$$

Fig. 3.5 Diffraction pattern of the compound $YBa_2Cu_3O_{6+\delta}$ **a** orthorhombic, **b** quadratic

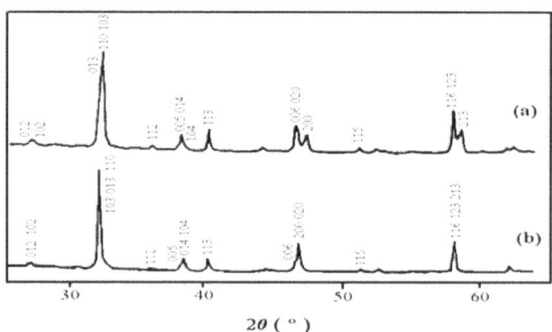

where

- N_j: Number of atoms in the crystal.
- f_j: Scattering factor of atom j.
- $exp(-B_j(\sin\theta/\lambda)^2)$: Fourier transform of the Gaussian probability cloud replacing the atom's point position, where B_j is the isotropic atomic displacement factor (or thermal agitation factor) of atom j.
- $exp(2i\pi(hx_j + ky_j + lz_j))$: Phase of the scattered waves with x_j, y_j, z_j being the reduced coordinates of atom j and hkl the Miller index identifying a family of lattice planes.

By definition, the diffracted intensity is $I_{hkl} = |F_{hkl}|^2$. Hence, we need to go from I_{hkl} obtained experimentally to the atomic coordinates x_j, y_j, z_j in the unit cell. Practically, a major difficulty makes structure resolution more laborious: only the magnitude of the structure factor is accessible experimentally, the phase of F_{hkl} remains unknown, posing the phase indetermination problem (Fig. 3.5).

3.3.4 Powder Diagram Technique

Atomic planes (hkl) of a crystal are randomly oriented relative to the monochromatic X-ray beam. θ is varied using a goniometer. Each time Bragg's Law is satisfied, we will witness constructive interferences and observe a reflection (hkl) of intensity $I(hkl)$. Indeed, Bragg's law links the properties of reciprocal space (the space of θ) and those of direct space (that of d). Each reflection is associated with a series of three indices h, k, l related to the parameters of the unit cell in direct space. For example, knowing the crystal parameters can provide a good idea of a ceramic's composition. For $YBa_2Cu_3O_{6+\delta}$, the calibration curves $a, b, c = f(\delta)$ are now well established. The most sensitive parameter is the c parameter, which provides a good precision value for δ.

The diffraction pattern provides at least two additional types of information:

- It informs about the purity of the examined material with the presence of parasitic phases (Y_2BaCuO_5, $Y_2Cu_2O_5$, $BaCuO_2$, CuO) indicated by reflections that do not belong to the studied crystal system.
- On the degree of organization of the material, indeed, theory predicts that the width of the reflections in reciprocal space, measured by $\Delta\theta$ at half-height, is inversely proportional to the correlation length in direct space.

3.3.5 Structure Refinement by Rietveld Method

X-ray diffraction is a potent technique as it provides structural and microstructural information about crystalline compounds. Currently, the Rietveld technique is utilized to solve structures from crystalline powders. It is not suitable for resolving large structures (more than ten atoms). However, this only applies to crystalline compounds. This method was proposed in 1969 by the Dutch crystallographer Hugo Rietveld [6].

The Rietveld method is based on the statistical principle of least squares, where the quantities to minimize are

$$M = \sum_{i=1}^{n} w_i (y_{i,\mathrm{obs}} - y_{i,c})^2 \quad \text{with} \quad w_i = 1/\sigma_i^2 \tag{3.5}$$

and

$$\chi^2 = \sum w_i [y(2\theta_i)_i - y(2\theta_i)_{ci}]^2 \tag{3.6}$$

- n: Number of observations.
- i: Step number measurement (corresponding to the value $(2\theta)_i$, i varies from 1 to n).
- w_i: The weight assigned to $y_{i,\mathrm{obs}}$.
- σ_i: Standard deviation on y_i (in counting statistics, the variance $\sigma_i = \sqrt{y_i}$).
- $y_{i,\mathrm{obs}}$: Observed intensity for the angle $2\theta_i$.
- $y_{i,c}$: Calculated intensity for the angle $2\theta_i$.

Minimizing the function M allows obtaining the positions of atoms in the lattice, as well as the associated atomic displacement parameters (described isotropically or anisotropically). It is also possible to add the occupancy rates of atoms on their crystallographic sites.

   The degree of convergence of a numerical refinement between the profile of observed intensity lines and that of calculated intensities is quantified using various Reliability (R) factors, classified into two categories.

Profile Agreement Factors

- Unweighted profile factor:

$$R_p = \frac{\sum_i |y_{i,\text{obs}} - y_{i,c}|}{\sum_i y_{i,\text{obs}}} \times 100 \tag{3.7}$$

- Weighted profile factor:

$$R_{wp} = \sqrt{\frac{\sum_i \omega_i (y_{i,\text{obs}} - y_{i,c})^2}{\sum_i \omega_i y_{i,\text{obs}}^2}} \times 100 \tag{3.8}$$

- GoF factor:

$$GoF = \sqrt{\frac{\sum_i \omega_i (y_{i,\text{obs}} - y_{i,c})^2}{N - P + C}} \tag{3.9}$$

   The weighted factor R_{wp} and the GoF (goodness of fit) factor best reflect the refinement progress because their numerator contains the residue χ^2 that is minimized.

$$\chi^2 = \left(\frac{R_{wp}}{R_{exp}}\right)^2 = \frac{M}{N - P + C} \quad \text{with} \quad R_{exp} = \sqrt{\frac{N - P + C}{\sum_i \omega_i y_{i,\text{obs}}^2}} = \frac{R_{wp}}{\sqrt{\chi^2}} \tag{3.10}$$

Structural Model Factors

- Bragg factor:

$$R_B = \frac{\sum_i |I_{i,\text{obs}} - I_{i,c}|}{\sum_i I_{i,\text{abs}}} \times 100 \tag{3.11}$$

- Structure factor:

$$R_F = \frac{\sum_i |\sqrt{I_{i,\text{obs}}} - \sqrt{I_{i,c}}|}{\sum_i \sqrt{I_{i,\text{obs}}}} \times 100 \tag{3.12}$$

⋄ M: Function to minimize.
⋄ N: Number of independent observations.
⋄ P: Number of adjusted parameters.
⋄ C: Number of constraints between the adjusted parameters.
⋄ $(N - P + C)$: Defines the number of degrees of freedom.

The statistical adjustment test noted as χ^2 should tend toward 1 for a successful refinement.

3.4 Iodometry

3.4.1 Definition

The oxygen content of superconducting materials significantly affects their structural, electrical, and magnetic properties. To precisely quantify this content, various analytical techniques are employed, including neutron diffraction, thermogravimetry, and iodometric titration. Among these, iodometry stands out for its simplicity, speed, and the availability of its chemical reagents [7].

The standard procedure involves grinding samples under controlled annealing conditions, then dissolving them in a diluted aqueous solution of hydrochloric acid with an appropriate amount of potassium iodide (KI) [8].

3.4.2 Chemical Reactions and Determination of Oxygen Content

For the analysis of the sample $YBa_2Cu_3O_{6+\delta}$, the key reactions are

Separation of ions:

$$YBa_2Cu_3O_{6+\delta} + 2(6 + \delta - x)HCl \rightarrow YCl_3 + 2BaCl_2 + 3CuCl_2 +$$
$$(2z - 1 - x)Cl^- + (6 + \delta - x)H_2O + xO \tag{3.13}$$

where x represents the number of oxygen atoms ($\frac{1}{2}O_2$) per mole.
In the presence of KI:

$$3CuCl_2 + 6KI \rightarrow 1.5Cu_2I_2 + 6KCl + 1.5I_2 \tag{3.14}$$

$$xO + 2xH^+ + 2xOH^- + 2xKI \rightarrow 2xKOH + xH_2O + xI_2 \tag{3.15}$$

$$\frac{1}{2}O_2 + H_2O + 2KI \rightarrow 2KOH + I_2 \tag{3.16}$$

Combining reactions 3.14 and 3.15 modifies reaction 3.13 to include the iodine produced, resulting in a comprehensive equation for the dissolution and reaction process in the presence of KI.

Addition of amidon:

When starch is added to the solution, it forms a dark blue complex with I_2, serving as an indicator for the completion of the chemical reaction, the disappearance of the dark blue color signals that equivalence has been reached.

Titration:

The quantity of I_2 molecules in the solution can be determined by titration with sodium thiosulfate ($Na_2S_2O_3$):

$$I_2 + 2Na_2S_2O_3 \rightarrow Na_2S_4O_6 + 2NaI \tag{3.17}$$

For $(1.5 + x)$ moles of I_2, we have

$$(1.5 + x)I_2 + (3 + 2x)Na_2S_2O_3 \rightarrow (1.5 + x)Na_2S_4O_6 + (3 + 2x)NaI \tag{3.18}$$

From reaction 3.13 we have

$$n_{123} = \frac{n_{I_2}}{(1.5 + x)} \tag{3.19}$$

From reaction 3.18 we have

$$\frac{n_{I_2}}{(1.5 + x)} = \frac{n_{Na}}{(3 + 2x)} \tag{3.20}$$

Combining reactions 3.19 and 3.20 we find:

$$3 + 2x = \frac{n_{Na}}{n_{123}} \tag{3.21}$$

where n_{Na} and n_{123} are the molar quantities of $Na_2S_2O_3$ and $YBa_2Cu_3O_{6+\delta}$, respectively:

$$n_{Na} = \frac{m_{Na}.V}{M_{Na}.V_{tot}} \quad \text{and} \quad n_{123} = \frac{m_{123}}{M_{123}} \tag{3.22}$$

where

- m_{123}: Mass of the used YBaCuO.
- M_{123}: Molar mass of $YBa_2Cu_3O_{6+\delta}$.
- m_{Na}: Mass of $Na_2S_2O_3$ used.
- M_{Na}: Molar mass of $Na_2S_2O_3$ (depends on the product, if it is $Na_2S_2O_3-5H_2O$, the mass equals 248.17 g).

- V_{tot}: Total volume of $Na_2S_2O_3$ (ml).
- V: Volume of the $Na_2S_2O_3$ solution used for the total titration (ml).

By substituting all parameters with their expressions in the equation for x, we obtain

$$x = \left(\frac{m_{123} \cdot V \cdot M_{(6+\delta)}}{M_{Na} \cdot V_{tot} \cdot m_{123}} - 3 \right) / \left(2 - \frac{16 \cdot m_{Na} \cdot V}{M_{Na} \cdot V_{tot} \cdot m_{123}} \right) \qquad (3.23)$$

Operational Mode: Regarding the procedure, we take the finely ground YBaCuO powder and place it in a beaker. We add the KI solution and then the HCl solution. The resulting solution should be brown in color with a precipitate. We add the starch solution and slowly mix the resulting solution using a magnetic stirrer, adding the $Na_2S_2O_3$ solution drop by drop using a graduated pipette until the mixture turns white. To ensure the titration is complete, we wait a few minutes without adding any $Na_2S_2O_3$. If the color turns back to brown, we add more, otherwise, the titration is complete. For samples doped with Sr and (Eu, Sm, or Nd), the mass of these elements is, of course, taken into account.

3.5 Measurement of Alternating Magnetic Susceptibility

3.5.1 Introduction

Magnetic susceptibility, denoted by χ, quantifies the response of a material to a magnetic excitation. It is defined by the equation $\chi = \frac{M}{H}$, where

- M is magnetization and
- H is the applied magnetic field.

The measurement of χ varies depending on the type of applied magnetic field:

- For a continuous or constant field, we measure the DC susceptibility (χ_{DC}).
- For an alternating field with constant amplitude and frequency, we measure the AC susceptibility (χ_{AC}).

In the case of AC susceptibility, an alternating magnetic field $H_{ac} = H_{ac}^0 \cos(\omega t)$ is applied, and the sample's response is measured using a detection coil. Thus, we have

$$\chi_{ac} = \frac{dM}{dH_{ac}} \qquad (3.24)$$

The measured susceptibility decomposes into two parts: $\chi = \chi' + i\chi''$. The real term χ' represents the Meissner effect, while the imaginary term χ'' is associated with dissipative phenomena.

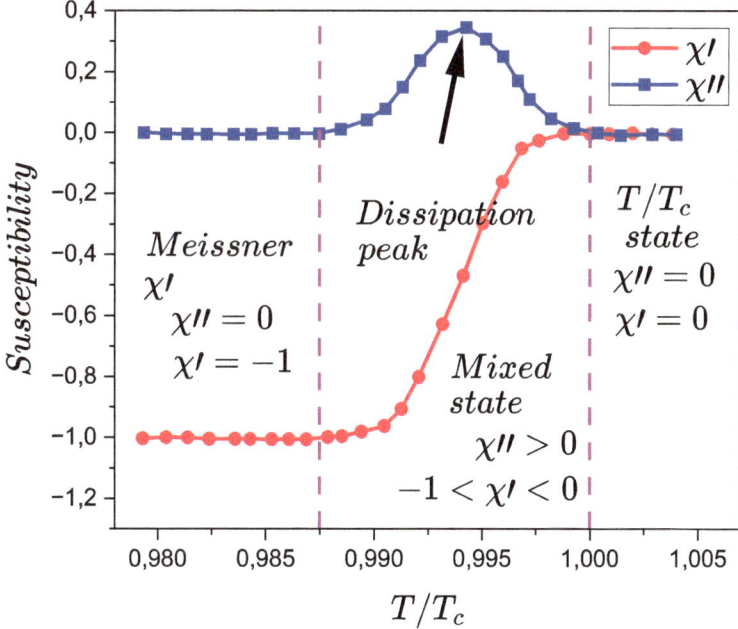

Fig. 3.6 Variation of χ' and χ'' of AC susceptibility as a function of reduced temperature

Gradually cooling a material subjected to an alternating magnetic field reveals a sharp drop in χ' at the critical temperature T_c, and χ'' reaches a maximum. Figure 3.6 illustrates the variation of AC susceptibility with temperature, marking the different states of the material during the normal/superconducting transition.

3.5.2 Measurement Principle

The total magnetic field applied to the sample varies with time, described by

$$H(t) = H_{dc} + H_{ac}^0\, e^{i\omega t} \quad \text{with} \quad H_{ac}^0 << H_{dc} \tag{3.25}$$

Susceptibility is expressed in complex form:

$$\chi = \chi' + i\chi'' = \frac{dM}{dH} = \frac{1}{\mu_0}\frac{dB}{dH} - 1 \tag{3.26}$$

The local average magnetic field in the sample is

$$\bar{B}(t) = \mu H(t) = (\mu' + i\mu'')H(t) \tag{3.27}$$

where $\mu = \chi + 1$ represents the complex permeability.

Multiplying successively by $\cos(\omega t)$ and $\sin(\omega t)$ and integrating over time from 0 to $2\pi/\omega$, we obtain

$$\chi' = \mu' - 1 = \frac{\omega}{\pi H_{ac}^0} \int_0^{2\pi/\omega} \cos(\omega t)\bar{B}(t)dt - 1 \tag{3.28}$$

$$\chi'' = \mu'' = \frac{\omega}{\pi H_{ac}^0} \int_0^{2\pi/\omega} \sin(\omega t)\bar{B}(t)dt \tag{3.29}$$

- In a completely superconducting state (Meissner state), the screening is perfect ($\chi' = -1$), and vortices are anchored by material defects ($\chi'' = 0$).
- As the temperature increases or the amplitude of the applied magnetic field becomes sufficient to move vortices, the screening diminishes ($-1 < \chi' < 0$), and energy dissipation appears ($\chi'' > 0$).
- In the normal state, the magnetic screening disappears ($\chi' = 0$), and flux lines move freely ($\chi'' = 0$).

The principle of measuring AC susceptibility involves subjecting the sample to a weak alternating field generated by a primary coil (transmitter). The sample's response to this magnetic excitation is measured by the secondary coil (receiver), known as the detection coil, which consists of two coils wound in opposite directions and connected in series, converting the flux passing through the sample into an electrical signal (Fig. 3.7) [9]. Indeed, the magnetic field produced by the primary coil induces opposing currents in the two halves of the secondary winding: the result is zero if the symmetry is perfect. In the case of the sample's presence in one of the halves of the secondary coil, it captures the voltage induced by the flux variation which is represented as

$$V(t) = -L\frac{d\phi}{dt} \approx -\frac{dB(t)}{dt} = V_0(\chi' \sin(\omega t) + \chi'' \cos(\omega t)) \tag{3.30}$$

- ϕ: The total magnetic flux seen by the detection coil and
- L: The mutual inductance between the primary coil and the detection coil.

$V(t)$ is measured by a mutual inductance bridge (Fig. 3.7) and sent to a dual synchronous detection that separates χ' and χ'' [10].

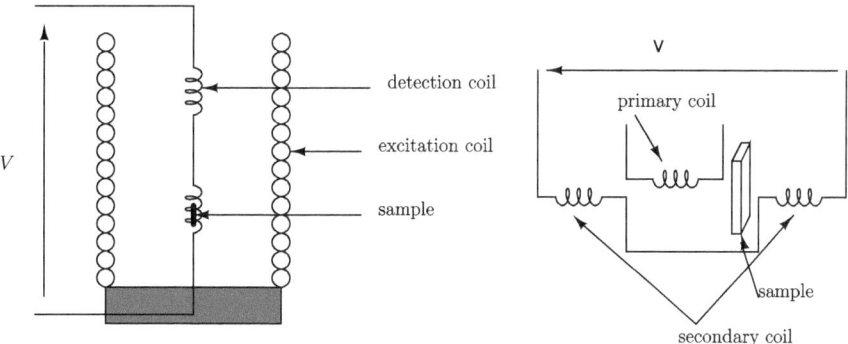

Fig. 3.7 Principle of assembly of the primary and secondary coil: **a** Positioning of the sample (powder or piece), **b** Positioning of the sample (pellets and thin films)

3.5.3 Experimental Setup

The experimental apparatus consists of a probe and a measurement system (Fig. 3.8).

3.5.3.1 Probe

It comprises three flat coils: one emitting (primary) coil and two receiving (secondary) coils placed symmetrically with respect to the emitting coil and connected in opposition. For measurements, the sample is placed between the emitting coil and the (+) secondary coil (Fig. 3.8).

- When $T > T_c$, both coils capture the same flux, and the mutual inductance is null.
- When $T < T_c$, the sample excludes flux via diamagnetic effect, and the mutual inductance coefficient between the coils changes significantly.

Note that in alternating regime, the flux exclusion is due to supercurrents induced by $H_{ac}(t)$, this is not the Meissner effect.

3.5.3.2 Thermometry

Copper wire braids ensure thermal contact between the sample and the rest of the device. Two temperature measurement resistors are placed symmetrically around the sample, and a heating wire is also wound symmetrically around the sample (Fig. 3.8a). For $4.2\,K < T < 20\,K$, a "tail" with low thermal conduction ensures an adjustable thermal leak. For $20\,K < T < 300\,K$, the temperature is regulated by the thermal gradient within the cryostat, adjusting the probe's height relative to the helium level. Fine temperature adjustment is achieved by injecting an adjustable current into the

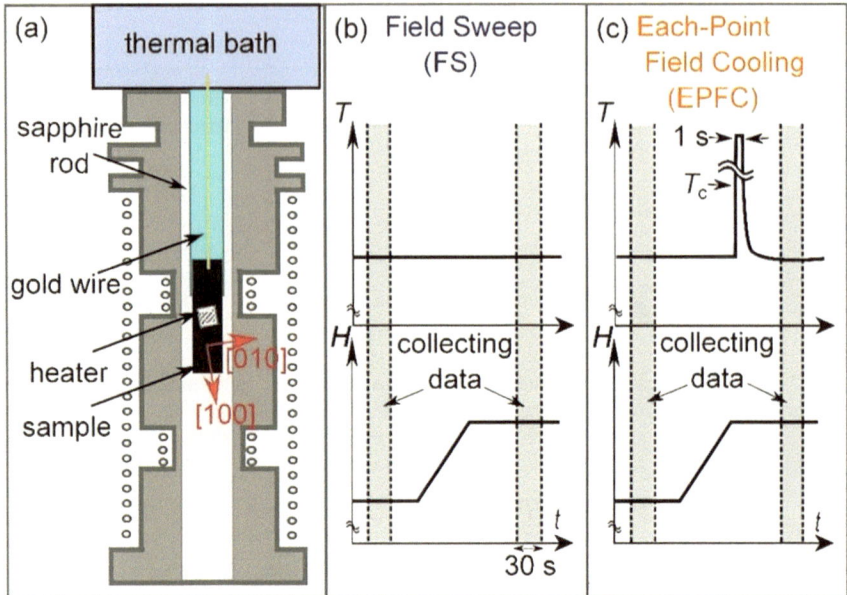

Fig. 3.8 Schematic of the sample mounted in an AC susceptometer coil [11]

heating wire. The desired temperature range is swept through slowly, with a minimal step of 0.2 K controlled by vertical movement of the rod.

3.5.3.3 Mutual Inductance Bridge

The principle of the mutual inductance bridge used is outlined in Fig. 3.9. The measurement principle is as follows: A generator injects a current I_p into the primary winding of a measurement probe. This current creates a magnetic field in the secondary, consisting of two identical coils connected in phase opposition, such that the mutual induction coefficient between primary and secondary is null. Introducing a magnetic sample into one of the secondary coils results in a mutual Me between primary and secondary, proportional to the complex susceptibility, in the form $\chi = \chi' - i\chi''$. Thus, the mutual coefficient is also complex: $Me = M'e - iM''e$. The unbalance voltage is in the form:

$$V_e = i M_e \omega I_p = (i M_e' + M_e'')\omega I_p \qquad (3.31)$$

The in-phase component behaves like a physical mutual, i.e., $i M_e' I_p$. The quadrature component $M_e'' I_p$ is resistive but varies proportionally to frequency.

The balance of the in-phase component is achieved by attenuating, with a factor varying from 0 to 1 in steps of 10^{-5}, the voltage provided by a reference mutual

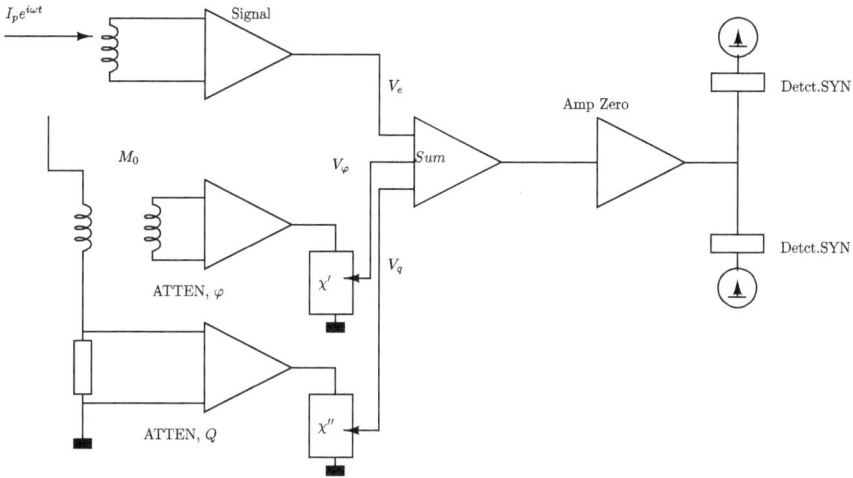

Fig. 3.9 Mutual inductance bridge

inductance M_0 of 1 mH. This realizes a variable mutual inductance from 0 to 1 mH with a resolution of 10^{-8}. The balance of the quadrature component is ensured by the voltage drop across a resistor R_0.

3.5.4 Measurement Conditions and Relevant Parameters

The measurements of alternating susceptibility (χ', χ'') were conducted under an alternating magnetic field $H_{ac} = 0.11$ Oe (frequency = 1500 Hz) and $0 < H_{dc} \leq 126$ Oe, initially without an external static H_{dc} field, then with varying fields, while heating the sample.

The curve in Fig. 3.10 illustrates the following physical characteristics:

- T_c: Critical temperature determined by the tangent to $\chi'(T)$ at the inflection point.
- T_{conset}: Critical onset temperature at the start of the transition from the normal state to the superconducting state.
- T_p: Temperature of the peak of $\chi''(T)$, corresponding to the onset of irreversibility.
- ΔT_c: Width of the superconductor-normal transition determined between temperatures corresponding to 10% and 90% of the height of $\chi'(T)$.
- ΔT_p: Half-width of the peak of $\chi''(T)$, indicating the behavior of grains under an external magnetic field.

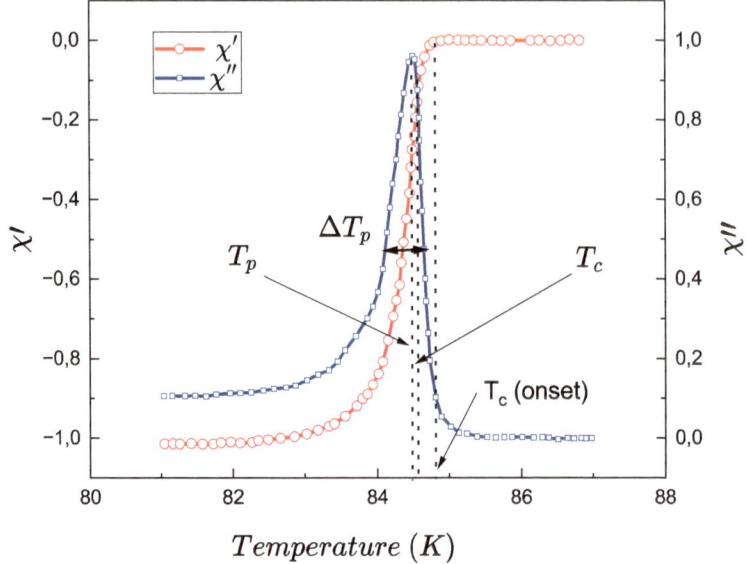

Fig. 3.10 Relevant parameters of χ_{ac}

3.6 Measurement of Electrical Resistivity

In the absence of an external field, the electrical resistance of a superconducting material becomes zero below T_c. The resistivity curve as a function of temperature, measured from room temperature to very low temperature, allows for the determination of the sample's critical temperature, characterizing the transition between the normal state and the superconducting state. It features five characteristic zones (Fig. 3.11):

- $T < T_{c\,offset}$: The sample is in the superconducting (S) state, thus $\rho(T)$ is null.
- $T = T_{c\,offset}$: Start of the resistive N-S transition.
- $T_{c\,offset} < T < T_{c\,onset}$: The sample is in the mixed or vortex state. A transition width S-N: $\Delta T = T_{c\,onset} - T_{c\,offset}$ is observed between temperatures corresponding to 10% and 90% of the height. This value characterizes the quality of the coupling between grains, for better current passage between grains, this width should be small, hence, the transition width qualifies the sample.
  Note: ΔT is also sensitive to crystal imperfections (impurities, grain boundaries, etc.).
- $T = T_{c\,onset}$: Disappearance of superconductivity.
- $T > T_{c\,onset}$: Metallic or normal behavior of the sample (ohmic regime), with $\rho = \rho_0 + \alpha T$ where ρ_0 is the residual resistivity (measurement noise, impurities...) and α is the slope of $\rho = \rho(T)$ representing the coupling of normal electrons–phonons.

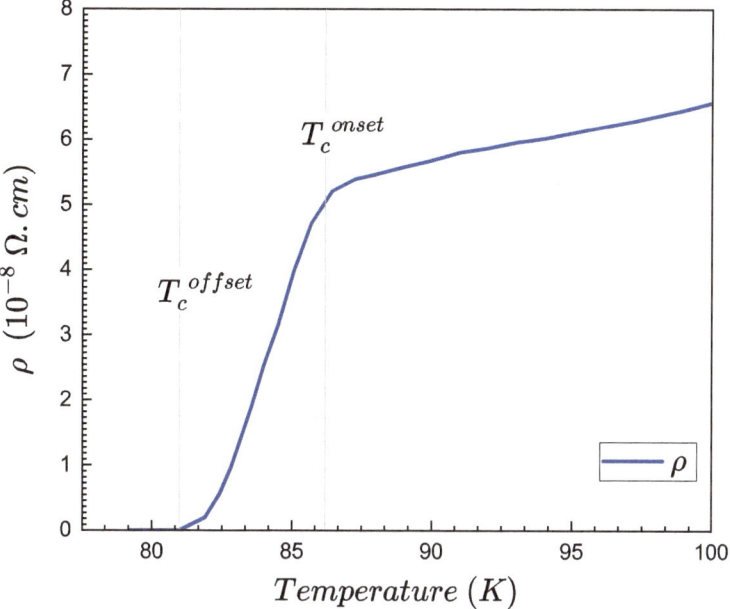

Fig. 3.11 Variation of resistivity as a function of temperature for an HTSC [12]

The most commonly used technique is the four-point or four-wire method: Two wires to inject current into the sample and two wires for measuring the voltage between two points along the current path.

3.6.1 Van der Pauw Method

The van der Pauw method can be applied to any sample provided the following rules are adhered to [13]:

- The sample must be flat.
- The sample must be homogeneous and compact, of any shape, where the thickness (d) is small relative to the lateral dimensions.
- The sample must not have isolated holes.
- The four metallic contacts should be placed on the edges of the sample in a symmetrical fashion (Fig. 3.12).

By applying an electrical current I (1 mA–10 mA) between contacts 1 and 2, and measuring the voltage between the other two contacts 3 and 4 using a microvoltmeter, a resistance $R_1 = V_{34}/I_{12}$ is obtained. To accurately determine the sample's resis-

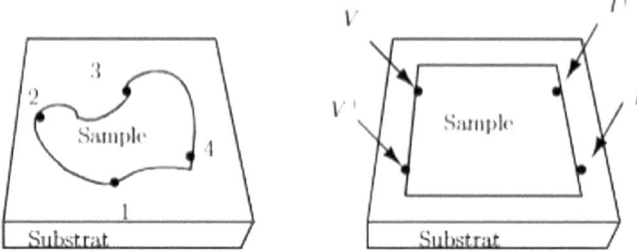

Fig. 3.12 Measurement schema for the van der Pauw method

tivity, resistivities $R_2 = V_{41}/I_{23}$, $R_3 = V_{12}/I_{34}$, and $R_4 = V_{23}/I_{41}$ are also measured through a simple cyclic rotation. The resistivity is given by

$$\rho = \frac{\pi}{\ln(2)} \bar{R} \times d \times F(R_1/R_2) \qquad (3.32)$$

where $\bar{R} = \frac{1}{4}\sum_{i=1}^{4} R_i$, d is the thickness, and $F(R_1/R_2)$ is a correction factor ranging between 0 and 1 that accounts for the asymmetrical aspect of the sample. Generally, for a symmetrically shaped sample (square, circular, or cross-shaped) $R_i \approx R_j$ so $R_1/R_2 \approx 1$ and the F factor equals 1. A significant correction is necessary if the R_1/R_2 ratio is greater or equal to 10. The van der Pauw method is commonly applied to the measurement of resistivity of thin layers.

3.6.2 Experimental Device and Procedure

The measurement of resistivity as a function of temperature is conducted in an apparatus that allows for the control and measurement of temperature. For this, the setup outlined in Fig. 3.13 is used.

The experimental setup consists of: a cryostat equipped with a cryogenic pump and a sample holder for measuring resistivity. The sample is placed in a vacuum insulation of the cryostat. A temperature controller allows for the regulation and reading of the sample's temperature (4 K–300 K). Wiring connects the four contacts of the sample to the contact box (voltage, current). The apparatus is currently used with liquid helium at temperatures ($4.2\,\text{K} < T < 300\,\text{K}$) controlled by the temperature regulator, using a carbon resistor ($4.2\,\text{K} < T < 60\,\text{K}$) and a platinum resistor ($60\,\text{K} < T < 300\,\text{K}$) (or a thermocouple) placed below the sample holder. The recording of measurement points (voltage, current, temperature) is performed by a microcomputer thanks to a data acquisition card (V, I, T) on the PC which gives $\rho = \rho(T)$.

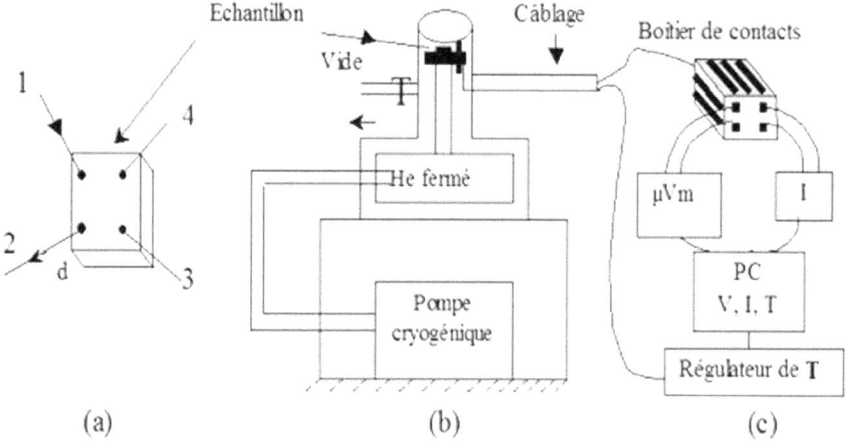

Fig. 3.13 Setup for measuring electrical resistivity

3.7 Conclusion

The chapter on experimental techniques provides an in-depth exploration of various methods employed in the preparation and characterization of superconducting materials. It outlines fabrication techniques such as spin coating, sputtering, and laser ablation for thin film deposition, as well as solid-state reactions for producing ceramic superconductors. The focus on polycrystalline samples $Y_{0.5}Ln_{0.5}SrBaCu_3O_{6+\delta}$ (Ln = Eu, Sm, and Nd) showcases a detailed synthesis process involving mixing, grinding, calcination, compaction, and sintering.

Furthermore, the chapter delves into characterization techniques such as X-ray diffraction, iodometry, measurement of alternating magnetic susceptibility, and electrical resistivity. X-ray diffraction elucidates the crystalline structure and composition of materials, while iodometry provides a simple yet effective method for determining oxygen content. Measurement of alternating magnetic susceptibility offers insights into the superconducting transition, and electrical resistivity measurements track the transition from normal to superconducting states.

The comprehensive discussion of experimental techniques underscores the intricate processes involved in both preparation and characterization, essential for advancing our understanding and application of superconducting materials. These techniques serve as valuable tools for researchers in optimizing material properties and advancing the field of superconductivity.

References

1. A. Aydi, *Élaboration et caractérisations diélectriques de céramiques ferroélectriques et/ou relaxeur de formule $MSnO_3$-$NaNbO_3$(M=Ba, Ca)* (Université de Bordeaux I; Faculté des Sciences de Sfax, Theses, 2005)
2. X-Ray diffraction, September 2022
3. B.S. Saini, R. Kaur, *X-ray Diffraction* (Elsevier, 2021), pp. 85–141
4. P.W. Atkins, *Éléments de chimie physique* (De Boeck Supérieur, 1998)
5. M. Zhou, A. Sun, L. Sun, P. Han, Effect of the oxygen distribution and inhomogeneity on the superconducting properties of $YBa_2Cu_3O_{7-\delta}$ +$xmol$ % $Y_2Ba_4CuMoO_y$ superconductors. Supercond. Sci. Technol. **26**(10), 105006 (2013). (August)
6. H.M. Rietveld, A profile refinement method for nuclear and magnetic structures. J. Appl. Crystallogr. **2**(2), 65–71 (1969). (June)
7. N.J. Calos, J.S. Forrester, G.B. Schaffer, A crystallographic contribution to the mechanism of a mechanically induced solid state reaction. J. Solid State Chem. **122**(2), 273–280 (1996). (March)
8. Y. Maeda, T. Minemura, H. Andoh, Thermo-photometric study on phase transitions in sputter-deposited Ag-Zn alloy thin films. Jpn. J. Appl. Phys. **26**(7A), L1218 (1987). (July)
9. M.I. Youssif, A.A. Bahgat, I.A. Ali, AC magnetic susceptibility technique for the characterization of high temperature superconductors. Egypt. J. Solids **23**(2), 231–250 (2000)
10. P.H.E. Meijer, Kamerlingh onnes and the discovery of superconductivity. Am. J. Phys. **62**(12):1105–1108, December 1994
11. D. Shibata, H. Tanaka, S. Yonezawa, T. Nojima, Y. Maeno, Quenched metastable vortex states in Sr_2RuO. Phys. Rev. B **91**(10), (2015)
12. L. Zani, Influence du dopage sur les propriétés de transport et d'aimantation de rubans polycristallins de Bi 2212 : ancrage des vortex. Ph.D. thesis, Ecole polytechnique, 1998. Thèse de doctorat dirigée par Régnier, Pierre Sciences des matériaux Palaiseau
13. J.G. Webster, H. Eren (eds.), *Measurement, Instrumentation, and Sensors Handbook: Two-Volume Set* (CRC Press, 2018)

Chapter 4
Effects of Isovalent Substitutions and Argon Heat Treatment on the Structural and Superconducting Properties of $Y_{0.5}Ln_{0.5}SrBaCu_3O_{6+\delta}$

4.1 Introduction

The majority of extensive research efforts in solid-state physics have been directed toward studying the high T_c superconducting cuprates since 1986. It is well known that the physical properties of these compounds are closely linked to their preparation conditions. In high T_c, superconductivity is carried by positive charges (holes) on the strongly correlated CuO_2 planes. The electronic states of the CuO_2 planes are sensitive and influenced by structural, electrical, and magnetic characteristics, such as the size of the ionic radius, the applied magnetic field, valence, the distribution of surrounding ions, and the oxygen content. Moreover, all these characteristics are strongly influenced by the thermal treatments applied during the sample preparation. Particularly, when Ln is a rare earth element, such as Eu, Sm, and Nd, various effects are expected due to the large ionic radius of Ln ions compared to Y.

Among the most studied compounds, we can mention $LnBa_2Cu_3O_{6+\delta}$ (Ln = Y or lanthanides) for several reasons: on one hand, the electrical transport characteristics can be easily varied by doping the compound with substitution elements [1, 2] or by varying the oxygen content. For example, when $0 < \delta < 0.5$, the compound $YBa_2Cu_3O_{6+\delta}$ is a tetragonal insulator and antiferromagnetic, and for $0.5 < \delta < 1$, the compound is orthorhombic (p-type and a metal) and becomes superconducting at low temperature [3–6]. For each oxygen content $6 + \delta$, T_c increases with the size of the ionic radius r of Ln ions [7, 8]. On the other hand, these compounds have a relatively high critical temperature ($T_c \approx 90$ K) above the nitrogen liquefaction temperature [9–11].

Numerous studies have been conducted on Sr substitution in $Ln(Ba_{1-x}Sr_x)_2Cu_3O_{6+\delta}$ (where Ln = Eu, Sm, and Nd) [12–15]. These studies conclude that for each Ln, the critical temperature and orthorhombicity decrease with the increase of the Sr concentration x [12, 13, 16]. For each x, the crystal structure and T_c depend on the size of the ionic radius r of Ln ions [17–19].

In this article, the samples were synthesized at high temperature and subjected to the effect of two thermal treatments ([O] and [AO]) to obtain a greater solubility of the Ln ion in $Y_{0.5}Ln_{0.5}SrBaCu_3O_{6+\delta}$. The powder spectra of X-ray diffraction (XRD) and measurements of alternating magnetic susceptibility were performed. It is interesting to check if an isovalent substitution of Ba^{2+} by Sr^{2+} with a smaller ionic radius would modify some of the results discussed above when Y^{3+} is replaced by Ln^{3+} with a larger ionic radius. With the aim of studying the role of Yttrium and barium atomic planes and searching for the conditions and factors governing superconductivity in these compounds, we have studied the structural and superconducting properties of high critical temperature superconducting materials $Y_{0.5}Ln_{0.5}SrBaCu_3O_{6+\delta}$. Indeed, the correlations between these properties, with the influence of argon thermal treatment and the ionic size of Ln, will be discussed.

4.2 Experimental Techniques

The polycrystalline samples $Y_{0.5}Ln_{0.5}SrBaCu_3O_{6+\delta}$ with Ln = Eu, Sm, and Nd were synthesized by the solid-state reaction of the respective oxides and carbonates. The chemicals were of 99.999% purity, except for $BaCO_3$, which was 99.99% pure. The products Y_2O_3, Ln_2O_3, $SrCO_3$, $BaCO_3$, and CuO were weighed according to the composition ratio, well mixed, and calcined at 950 °C in air for a period of 12–18 h. The resulting product was ground, mixed, pelletized, and fired in air at 980 °C for a period of 16–24 h. This was repeated twice. The pellets were annealed in oxygen at 450 °C for a period of 60 h to 72 h and subsequently cooled. This was named the [O] sample for each Ln.

Powder X-ray diffraction data were collected at room temperature using a diffractometer equipped with a secondary beam graphite monochromator with a $CuK\alpha$ radiation source (40 kV/20 mA). The 2θ angle was varied from $20^circ irc$ to $120°$ in steps of $0.025°$, and the counting time per step was 10 s. The crystal structure was refined by Rietveld analysis of the powder X-ray diffraction data. For each sample, the oxygen content was close to 6.94 ± 0.04, determined by the iodometric method.

The magnetic response of the sample was detected by a pickup coil surrounding the sample. Measurements of the alternating magnetic susceptibility ($\chi_{ac} = \chi' + i\chi''$) as a function of temperature were performed at 1500 Hz in a field of 0.11 Oe. Furthermore, χ' and χ'' were measured in a static field ($0 < H = H_{dc} < 150$ Oe) superimposed on the alternating field of $Hac = 0.11$ Oe.

For each Ln, the same [O] sample was then heated under argon at 850 °C for about 12 h, cooled to 20 °C, and then the sample was annealed at 450 °C under oxygen for about 72 h. This sample was named the [AO] sample. XRD and measurements of the alternating susceptibility were performed on a part of this sample.

4.3 Results

Figure 4.1a and b displays the XRD powder patterns of the systems Ln = Eu, Sm, and Nd, respectively. The samples were well crystallized, and the reflections were sharper and better resolved after the [AO] thermal treatment. The lattice parameters determined from the XRD patterns by Rietveld refinement are collected in Table 5.1. These diffraction patterns confirmed that all samples have an orthorhombic perovskite structure, and no impurity phase was observed after the [AO] thermal treatment (indicated by a cross in Fig. 4.1a disappeared after the [AO] thermal treatment). This indicates an improvement in the crystallographic quality of the [AO] samples. As seen in Fig. 4.2, for each thermal treatment in $Y_{0.5}Ln_{0.5}SrBaCu_3O_{6+\delta}$, the c parameter decreases, b is almost constant for all systems but a increases with the ionic radius $r(Ln^{3+})$ for Ln = Eu to Sm, then decreases for Nd leading to a decrease in orthorhombicity $\epsilon = (b - a)/(b + a)$. This is due to the fact that the ionic radii of Eu^{3+} (0.95 0.95 Å), Sm^{3+} (0.965 Å), and Nd^{3+} (0.995 Å) ions are larger than that of Y^{3+} (0.893 Å). For each Ln, the [AO] thermal treatment increases b and c but decreases a. While the volume V of the unit cell remains constant indicating a rearrangement of the unit cell (Table 5.1). Moreover, the [AO] thermal treatment increases ϵ, for $Y_{0.5}Eu_{0.5}SrBaCu_3O_{6+\delta}$ from 5.872×10^{-3} to 8.355×10^{-3}, for $Y_{0.5}Sm_{0.5}SrBaCu_3O_{6+\delta}$ from 4.033×10^{-3} to 6.638×10^{-3}, and for $Y_{0.5}Nd_{0.5}SrBaCu_3O_{6+\delta}$ from 3.913×10^{-3} to 7.174×10^{-3} in Fig. 4.10.

To gain insight into the internal strains induced by Ln substitution, we calculated the interatomic distances $d_{Cu(1)-(Sr,Ba)}$. An increase in this distance is observed with $r(Ln^{3+})$, indicating an increase in the strain introduced by the larger ionic radius replacing the smaller Y atom in the unit cell. For each $r(Ln^{3+})$, the [AO] thermal treatment reduces the distance $d[Cu(1) - (Sr/Ba)]$ (Table 4.1).

The temperature dependence of the real part of the susceptibility (χ') is graphically represented in Fig. 4.3a, b for both [O] and [AO] thermal treatments. Since the same sample was used for both thermal treatments, one can compare the diamagnetic shielding response (amplitude $\chi'(T)$) and note that the anchoring current in the [AO] sample has significantly increased compared to the [O] sample. The superconducting

Table 4.1 T_c and crystalline parameters of $Y_{0.5}Ln_{0.5}BaSrCu_3O_{6+\delta}$ as a function of thermal treatment

Ln	r(Ln) (Å)	Tr. ther	a (Å)	b (Å)	c (Å)	V (Å³)	ϵ 0.10^{-3}	T_c (K)	$d_{Cu(1)-(Sr,Ba)}$ (Å)
Eu	0.950	[O]	3.81	3.85	11.58	170.02	5.872	82.1	3.43811
		[AO]	3.80	3.86	11.59	169.94	8.355	82.5	3.43759
Sm	0.965	[O]	3.83	3.86	11.58	171.06	4.033	81.0	3.43582
		[AO]	3.82	3.87	11.58	170.88	6.638	82.0	3.43491
Nd	0.995	[O]	3.82	3.85	11.56	169.82	3.913	78.9	3.43502
		[AO]	3.81	3.86	11.57	170.05	7.174	80.5	3.43173

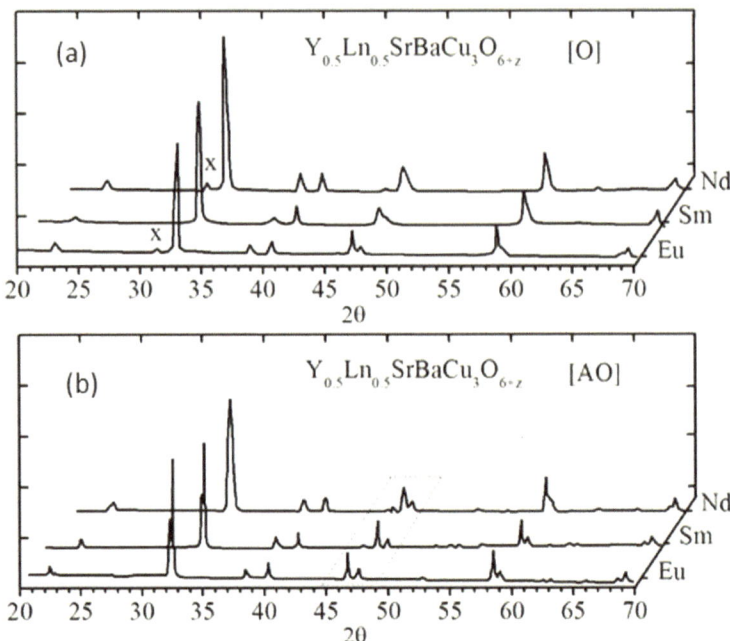

Fig. 4.1   XRD patterns of $Y_{0.5}Ln_{0.5}SrBaCu_3O_{6+\delta}$. **a** [O] samples, **b** [AO] samples [20, 21]

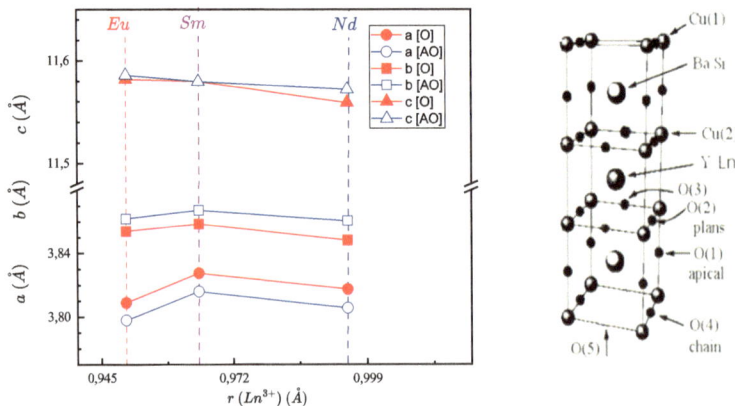

Fig. 4.2   (left) Crystalline parameters as a function of the ionic radius $r(Ln^{3+})$. (right) Unit cell of $Y_{0.5}Ln_{0.5}SrBaCu_3O_{6+\delta}$ [20, 21]

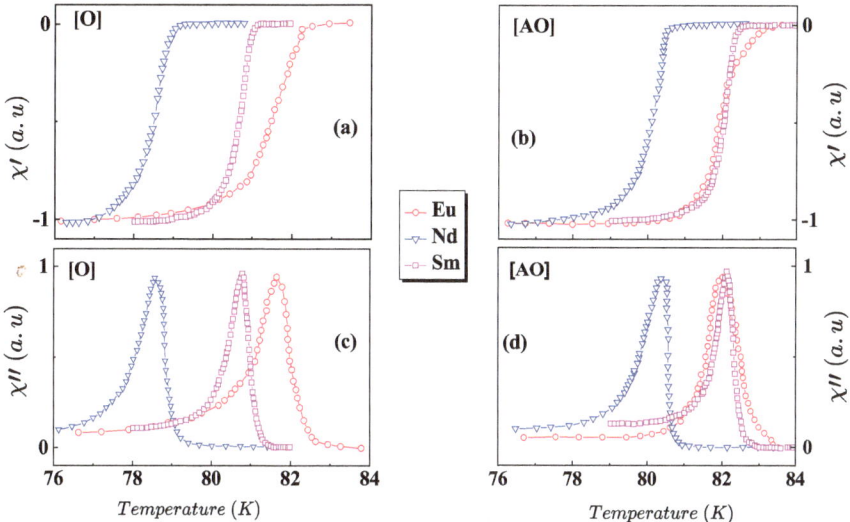

Fig. 4.3 χ' and χ'' of $Y_{0.5}Ln_{0.5}SrBaCu_3O_{6+\delta}$ as a function of temperature and thermal treatment: (a, c) [O] samples, (b, d) [AO] samples [20, 21]

Fig. 4.4 T_c as a function of $r(Ln^{3+})$ and thermal treatment of $Y_{0.5}Ln_{0.5}SrBaCu_3O_{6+\delta}$ with Ln = Eu, Sm, and Nd [20, 21]

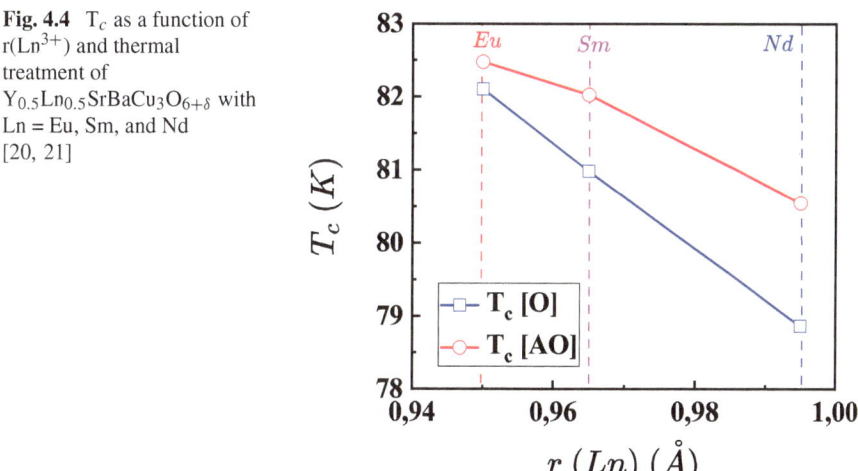

transition temperatures T_c, defined as the start of the diamagnetic transition, increase from 82.1 K [O], 80.98 K [O], and 78.86 K [O] to 82.47 K [AO], 82.02 K [AO], and 80.54 K [AO] for the Ln = Eu, Sm, and Nd samples, respectively.

In Fig. 4.4, we showed the dependence of T_c as a function of the ionic radius $r(Ln^{3+})$ of the $Y_{0.5}Ln_{0.5}SrBaCu_3O_{6+\delta}$ systems. For each Ln ionic radius, the [AO] thermal treatment increases T_c. For each thermal treatment, as the ionic size r of the rare earth ion decreases T_c increases until reaching its maximum value for Ln = Eu. A maximum increase of T_c of a value equal to $\delta T_c = T_c[AO] - T_c[O] = 1.68$ K in the Ln = Nd sample and a minimum of $\delta T_c = T_c[AO] - T_c[O] = 0.37\,\text{K}$ in the Ln = Eu sample were observed. Thus, T_c depends on the thermal treatment and the ionic radius of the rare earth.

4.3.1 Imaginary Part of the Alternating Magnetic Susceptibility and Irreversibility Line

From the results of the measurements of the imaginary part of the susceptibility (χ'') of the samples (Fig. 4.3c, d), it can be seen that χ'' is very sensitive to thermal treatment and the size of the Ln ionic radius. The T_p peak in χ'' reflects the intergranular critical current [22]. As the Ln ionic radius increases, p is shifted toward lower temperatures in the case of the [O] samples. For each Ln, T_p increased after the [AO] thermal treatment following the increase in T_c.

The static magnetic field $H_{dc}(t)$ sets an upper limit to the irreversibility line (L.I.) marking the beginning of dissipation and the region in which a superconductor can remain useful. In fact, when the static field H_{dc} is plotted as a function of $t = T_p/T_c$ (with $T_p = T_{irr}$) for the six samples (Fig. 4.5), an improvement (an increase in slope) of the irreversibility line was observed following the argon-oxygen thermal treatment.

These results can be analyzed using the following equation $H_{dc} = K'(1 - t)^n$ [23]. Straight lines were obtained when $\ln(H_{dc})$ was plotted as a function of $\ln(1 - t)$ in (Fig. 4.6). For each thermal treatment, K' remarkably increases with $r(Ln^{3+})$ (Fig. 4.7).

For each Ln the [AO] thermal treatment increases the field K' and n, indicating an improvement in the vortex pinning properties. For example, the value of K' was estimated at 6.982 KOe and 366.55 KOe, respectively, for the [O] and [AO] samples in $Y_{0.5}Nd_{0.5}SrBaCu_3O_{6+\delta}$ (Table 4.2). K' can be interpreted as the field necessary to reduce the intergranular critical current to zero in the limit of $T_p = 0$, and n defining the type of Josephson junction between grains in the superconductor.

Table 4.2 Superconducting parameters of $Y_{0.5}Ln_{0.5}SrBaCu_3O_{6+\delta}$ as a function of $r(Ln^{3+})$ and thermal treatment [20, 21]

Ln	$r(Ln^{3+})$ (Å)	Tr. ther	T_p (K)	K'(Oe)	n
Eu	0.950	[O]	81.6	1463.8	1.19
		[AO]	81.9	5860.0	1.55
Sm	0.965	[O]	80.8	2423.1	1.22
		[AO]	81.2	9011.6	1.86
Nd	0.995	[O]	78.6	6982.0	1.90
		[AO]	80.0	366553.6	2.76

4.4 Discussion

The samples were prepared in 1 atm of oxygen. Moreover, the [AO] thermal treatment did not significantly change the total oxygen content $6 + \delta$, which was about 6.94 $\pm$ 0.04 according to iodometric measurements, but it increased T_c. Thus, the reason for this increase can be explained by factors other than δ.

It is well known that $YBa_2Cu_3O_{6+\delta}$ materials are orthorhombic superconductors, and their T_c is close to 92 K [24]. These materials are characterized by double $Cu(2)O_2$ layers (oriented along the a-b plane) responsible for supercurrent transport and $Cu(1)O$ chains (along the b direction) that act as a charge reservoir for these planes [25, 26]. As can be seen in Table 4.1, the substitution with rare earths causes variations in the a, b, and c crystalline parameters but maintains the orthorhombic character, although there are minor changes in orthorhombicity. This indicates that the size of the ionic radius $r(Ln^{3+})$ does not alter the orthorhombic structure of the $Y_{0.5}Ln_{0.5}SrBaCu_3O_{6+\delta}$ systems. We found that the argon thermal treatment increased orthorhombicity, indicating an improvement in structural and superconducting properties. G. Uimin et al. indicate that this increase in orthorhombicity is due to increased oxygen order in the chains [27, 28]. These authors indicated that hole transfer in the $Cu(2)O_2$ planes is strongly related to the mechanism of oxygen filling in the basal

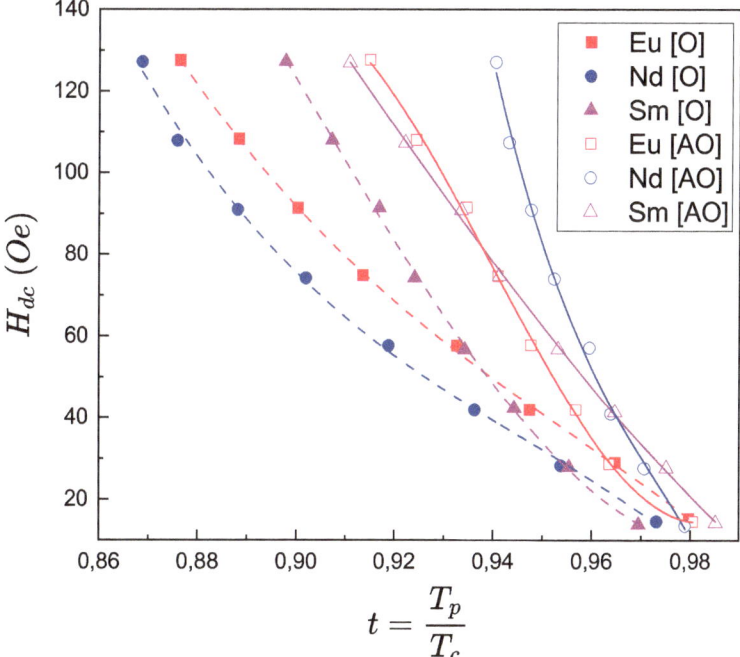

Fig. 4.5 H_{dc} as a function of $t = T_p/T_c$ and thermal treatment for $Y_{0.5}Ln_{0.5}SrBaCu_3O_{6+\delta}$ with Ln = Eu, Sm, and Nd [20, 21]

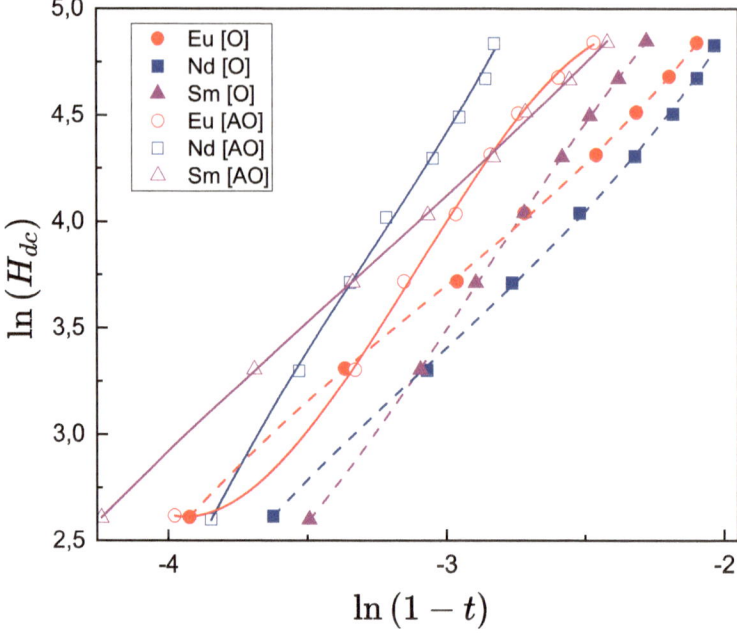

Fig. 4.6 $\ln(H_{dc})$ as a function of $\ln(1 - t)$ and thermal treatment of $Y_{0.5}Ln_{0.5}SrBaCu_3O_{6+\delta}$ with Ln = Eu, Sm, and Nd [20, 21]

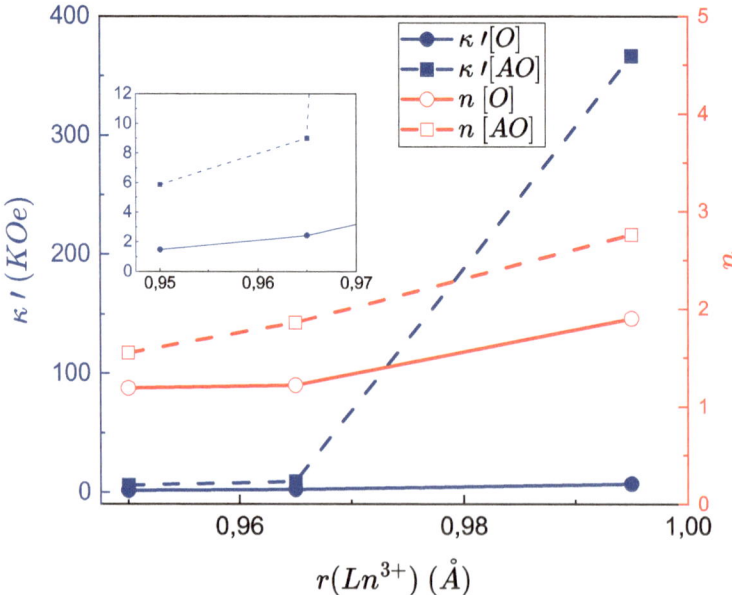

Fig. 4.7 Field K' as a function of the ionic radius $r(Ln^{3+})$ and thermal treatment of $Y_{0.5}Ln_{0.5}SrBaCu_3O_{6+\delta}$ with Ln = Eu, Sm, and Nd [20, 21]

plane at the O(4) and O(5) sites along the b- and a-axes, respectively, as shown in the unit cell (Fig. 4.2). This would lead to optimal superconducting properties and could explain the observed increase in T_c and the irreversibility line.

As shown in Fig. 4.2, for each Ln, the [AO] thermal treatment decreases the a crystalline parameter and increases b and c. This increase in the b parameter leads to an increase in the number of oxygen atoms per chain (NOC) along the b-axis (increase in anionic order in the basal plane). This indicates an improvement in hole transfer from the Cu(1)O chains (along the b direction) to the superconducting planes, via the apical oxygen O(1) (Fig. 4.2). This argument is justified by the decrease in the interatomic distance $d_{Cu(1)-(Sr,Ba)}$ (Fig. 4.9).

Consequently, the number of holes ρ (Fig. 4.8) in the superconducting Cu(2)O$_2$ planes increases (deduced from the saturation subzone of the universal relationship that exists between τc as a function of ρ (with $\tau_c = T_c / T_c^{max}$) [29]) and T_c. For each $r(\text{Ln}^{3+})$, the [AO] thermal treatment increases ϵ and T_c and decreases $d_{Cu(1)-(Sr,Ba)}$. These results indicate a rearrangement of the same volume V of the unit cell and improvement of charge transfer from the chains to the superconducting planes leading to an increase in T_c.

When the Ln ion occupies the Ba/Sr sites, the same amount of Ba/Sr cations is pushed into the Y plane (with Ln a trivalent ion). This increases the positive charge density around the Ba/Sr site and the attractive force with oxygens. Conse-

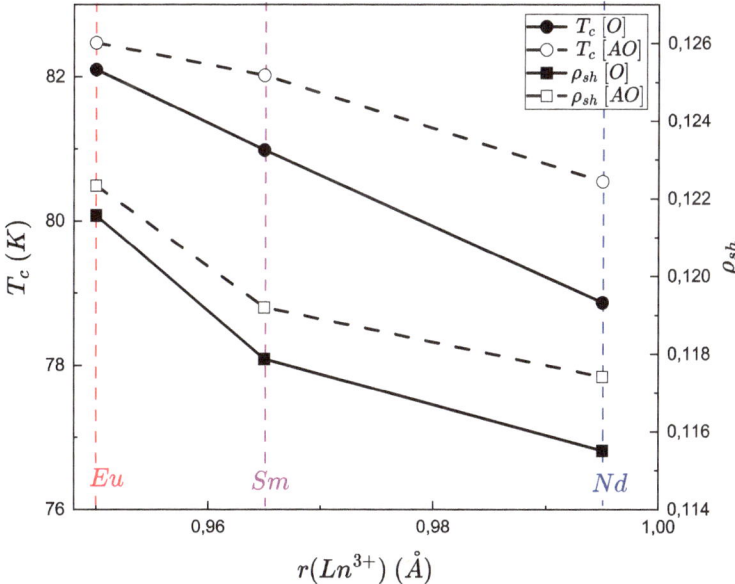

Fig. 4.8 Correlation between ρ and T_c as a function of the ionic radius $r(\text{Ln}^{3+})$ and thermal treatment of $Y_{0.5}Ln_{0.5}SrBaCu_3O_{6+\delta}$ with Ln = Eu, Sm, and Nd [20, 21]

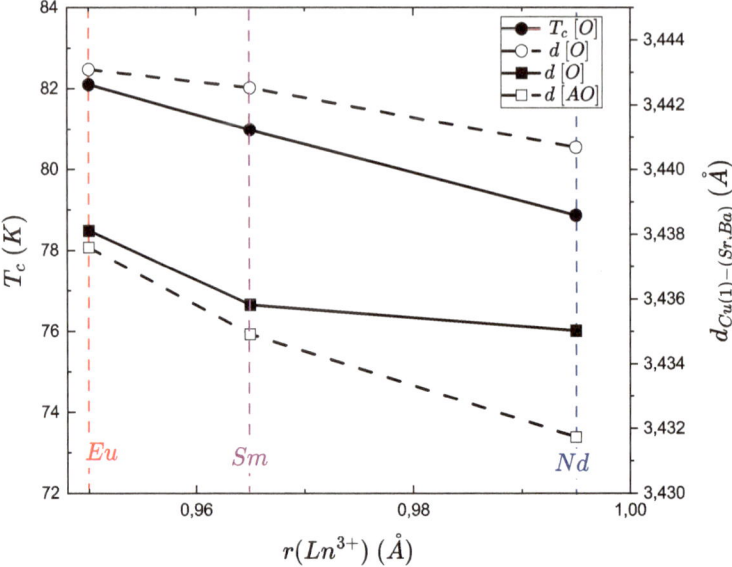

Fig. 4.9 Correlation between $d_{Cu(1)-(Sr,Ba)}$ and T_c as a function of the ionic radius $r(Ln^{3+})$ and thermal treatment of $Y_{0.5}Ln_{0.5}SrBaCu_3O_{6+\delta}$ with $Ln = Eu$, Sm, and Nd [20, 21]

quently, vacant oxygen sites O(4) have a greater chance of being filled. On the other hand, Ba^{2+}/Sr^{2+} occupying the Y^{3+}/Ln^{3+} site reduces the strength of the attractive force applied to oxygen in the $Cu(2)O_2$ plane. This increases the $Cu(2)–O(2)–Cu(2)$ distortion angle along the b-axis. Changes in the cationic sites increase b and decrease a after the [AO] thermal treatment. This indicates the movement of oxygens from the O(5) sites to the vacant O(4) sites along the b-axis. This increases NOC and the order of oxygen chains along the b direction. This NOC increase is responsible for the transfer of negative charges from the $Cu(2)O_2$ planes to the $Cu(1)$-O chains via O(1), thus increasing the number of holes ρ in the superconducting planes. For each $r(Ln^{3+})$, the [AO] thermal treatment increases ρ indicating an improvement in the superconducting properties in these systems.

Both arguments (cationic and anionic disorders) are justified here by the three remarkable correlations observed between $T_c(r)$, $\rho(r)$, and $d_{Cu(1)-(Sr,Ba)}(r)$ (Figs. 4.8 and 4.9), respectively, and, on the other hand, between $\Delta T_c(r) = T_c[AO] - T_c[O]$ and $\Delta\epsilon(r) = \epsilon[AO]-\epsilon[O]$ in Fig. 4.10. Thus, structural and superconducting properties are correlated with the effect of the argon thermal treatment. Consequently, we are tempted to believe that the changes (increase or decrease) observed in T_c should not be related only to the ionic size of the rare earth but rather to a combination of several factors such as changes in $Cu(1)–O(1)$ distances, anionic oxygen disorders, and hole density.

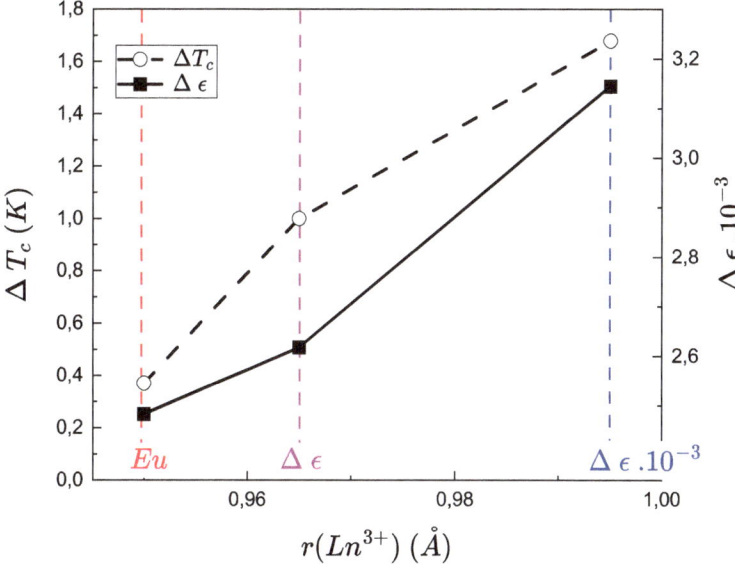

Fig. 4.10 Correlation between $\delta T_c = T_c[AO] - T_c[O]$ and $\delta\epsilon = \epsilon[AO] - \epsilon[O]$ as a function of the ionic radius $r(Ln^{3+})$ and thermal treatment of $Y_{0.5}Ln_{0.5}SrBaCu_3O_{6+\delta}$ with Ln = Eu, Sm, and Nd [20, 21]

4.5 Conclusion

In summary, the effect of Ln substitution on the structural and superconducting properties of the $Y_{0.5}Ln_{0.5}SrBaCu_3O_{6+\delta}$ system has been analyzed. To determine the structural properties, the results of the XRD measurements were analyzed using Rietveld refinement. The superconducting properties were analyzed using the measurement of alternating magnetic susceptibility. The following results were obtained:

- No evident impurity phase after the [AO] thermal treatment in the XRD results. This result shows an improvement in the crystallographic quality of the [AO] samples.
- For each thermal treatment, T_c increases as the ionic radius $r(Ln^{3+})$ decreases.
- For each size of the Ln ionic radius, the [AO] thermal treatment increases T_c. Regardless of $r(Ln^{3+})$ and thermal treatment, T_c increases as the ionic radius $r(Ln^{3+})$ decreases and reaches a maximum for Ln = Eu. A maximum increase in T_c of $\Delta T_c = 1.68$ K in the Ln = Nd sample and a minimum of $\Delta T_c = 0.37$ K in the Ln = Eu sample were observed. Therefore, T_c depends on the thermal treatment and the ionic radius of the rare earth.
- The observed improvement in the irreversibility lines of $Y_{0.5}Ln_{0.5}SrBaCu_3O_{6+\delta}$ after argon thermal treatment was explained by the improvement in grain quality, intergranular coupling, and crystallographic quality of the [AO] samples. Three

remarkable correlations were observed. Thus, the structural and superconducting properties are correlated with the effect of argon thermal treatment.

- The [AO] thermal treatment leads to an increase in the number of oxygen atoms per chain and decreases the distance between the chains and the $Cu(2)O_2$ superconducting planes, which increases the number of holes ρ. This improves charge transfer via $O(1)$, indicating an improvement in T_c and superconducting properties in the systems.

All these results stem from an interaction between the cationic disorder of Ln^{3+} on the Sr/Ba site along the c-axis and the anionic disorder of oxygens in the basal plane. A combination of several factors such as the ionic size of the rare earth, the decrease in the distance $d_{[Cu(1)-(Sr/Ba)]}$, the increase in cationic order and oxygen order in the chains, and the purity of the phase of the [AO] samples could explain the observed results.

References

1. H.A. Borges, M.A. Continentino, Pressure study of the paraconductivity of high T_c superconductors. Solid State Commun. **80**(3), 197–199 (1991)
2. R.V. Vovk, N.R. Vovk, G.. Ya.. Khadzhai, I.L. Goulatis, A. Chroneos, Effect of praseodymium on the electrical resistance of $YBa_2Cu_3O_{7-\delta}$ single crystals. Solid State Commun. **190**, 18–22 (2014)
3. S. Sadewasser, J.S. Schilling, A.P. Paulikas, B.W. Veal, Pressure dependence of T_c to 17 GPa with and without relaxation effects in superconducting $YBa_2Cu_3O_x$. Phys. Rev. B **61**(1), 741–749 (2000)
4. R. Beyers, B.T. Ahn, G. Gorman, V.Y. Lee, S.S.P. Parkin, M.L. Ramirez, K.P. Roche, J.E. Vazquez, T.M. Gür, R.A. Huggins, Oxygen ordering, phase separation and the 60 K and 90 K plateaus in $YBa_2Cu_3O_x$. Nature **340**(6235), 619–621 (1989)
5. Ruslan V. Vovk, Nikolaj R. Vovk, Oleksandr V. Dobrovolskiy, Effect of structural relaxation on the in-plane electrical resistance of oxygen-underdoped $REBa_2Cu_3O_{7-\delta}$ ((Re = Y, Ho) single crystals. J. Low Temp. Phys. **175**(3–4), 614–630 (2014)
6. R.J. Cava, B. Batlogg, C.H. Chen, E.A. Rietman, S.M. Zahurak, D. Werder, Oxygen stoichiometry, superconductivity and normal-state properties of $YBa_2Cu_3O_{7-\delta}$. Nature **329**(6138), 423–425 (1987)
7. G.V.M. Williams, J.L. Tallon, Ion size effects on 7 and interplanar coupling in $RBa_2Cu_3O_{7-\delta}$. Phys. C Supercond. **258**(1–2), 41–46 (1996)
8. J.G. Lin, C.Y. Huang, Y.Y. Xue, C.W. Chu, X.W. Cao, J.C. Ho, Origin of the R-ion effect on T_c in $RBa_2Cu_3O_7$. Phys. Rev. B **51**(18), 12900–12903 (1995)
9. M.K. Wu, J.R. Ashburn, C.J. Torng, P.H. Hor, R.L. Meng, L. Gao, Z.J. Huang, Y.Q. Wang, C.W. Chu, Superconductivity at 93 K in a new mixed-phase Y-Ba-Cu-O compound system at ambient pressure. Phys. Rev. Lett. **58**(9), 908–910 (1987)
10. B. Raveau, C. Michel, M. Hervieu, D. Groult, *Crystal Chemistry of High-T_c Superconducting Copper Oxides* (Springer, Berlin Heidelberg, 1991)
11. R.V. Vovk, N.R. Vovk, G.. Ya.. Khadzhai, I.L. Goulatis, A. Chroneos, Effect of high pressure on the electrical resistivity of optimally doped $YBa_2Cu_3O_{7-\delta}$ single crystals with unidirectional planar defects. Phys. B Condens. Matter **422**, 33–35 (2013)
12. R.A. Gunasekaran, B. Hellebrand, P.L. Steger, Crystal structure, oxygen stoichiometry and superconducting properties of $GdBa_{2-x}Sr_xCu_3O_{7-\delta}$ ($0.0 \leq x \leq 1.6$). Phys. C Supercond. **270**(1–2), 25–34 (1996)

13. X.Z. Wang, D. Bäuerle, High-T_c superconductivity in $GdBaSrCu_3O_7$. Phys. C Supercond. **176**(4–6), 507–510 (1991)

14. Y.G. Zhao, S.Y. Xiong, Y.P. Li, B. Zhang, S.S. Fan, B. Yin, J.W. Li, S.Q. Guo, W.H. Tang, G.H. Rao, D.J. Dong, B.S. Cao, B.L. Gu, Structural, transport, and magnetic properties of $PrBa_{2-x}Sr_xCu_3O_{7-\delta}$. Phys. Rev. B **56**(14), 9153–9157 (1997)

15. Tuerxun Wuernisha, Yumiko Takahashi, Kouichi Takase, Yoshiki Takano, Kazuko Sekizawa, Effects of sr substitution for ba on the crystal structure, oxygen content and electrical properties of $Nd(Ba_{1-x}Sr_x)_2Cu_3O_{7-\delta}$. J. Alloy. Compd. **377**(1–2), 216–220 (2004)

16. P. Karen, H. Fjellvåg, A. Kjekshus, A.F. Andresen, Chemical pressure and other effects of strontium substitution in $YBa_2Cu_3O_{9-\delta}$. J. Solid State Chem. **92**(1), 57–67 (1991)

17. X.Z. Wang, B. Hellebrand, D. Bäuerle, Crystal structure and superconductivity in REBaSrCu$_3$O$_x$. Phys. C Supercond. **200**(1–2), 12–16 (1992)

18. X.Z. Wang, B. Hellebrand, D. Bäuerle, M. Strecker, G. Wortmann, W. Lang, Oxygen ordering and superconductivity in $GdBaSrCu_3O_{7-x}$. Phys. C Supercond. **242**(1–2), 55–62 (1995)

19. X.Z. Wang, P.L. Steger, M. Reissner, W. Steiner, Oxygen ordering and superconductivity in $DyBaSrCu_3O_y$. Phys. C Supercond. **196**(3–4), 247–251 (1992)

20. K. Khallouq, A. Nafidi, A. Aboulkassim, M. Bellioua, E.Y. El Yakoubi, A. Khalal, Effects of isovalents substitutions and argon heat treatment on the structural and superconducting properties of $Y_{0.5}Ln_{0.5}SrBaCu_3O_{6+z}$. Int. J. Eng. Res. Technol. **5**(10), 179, 10 2016. Total Downloads: 179

21. K. Khallouq, A. Nafidi, A. Aboulkassim, E.Y. El Yakoubi, E.-S. Es saïd, Effects of isovalent substitutions and heat treatment on structural and superconducting properties of high-critical temperature superconductors. Mater. Today Proc. **22**, 140–145 (2020)

22. H. Küpfer, I. Apfelstedt, R. Flükiger, C. Keller, R. Meier-Hirmer, B. Runtsch, A. Turowski, U. Wiech, T. Wolf, Intragrain junctions in $YBa_2Cu_3O_{7-x}$ ceramics and single crystals. Cryogenics **29**(3), 268–280 (1989)

23. K.A. Müller, M. Takashige, J.G. Bednorz, Flux trapping and superconductive glass state in $La_2CuO_{4-y}Ba$. Phys. Rev. Lett. **58**(11), 1143–1146 (1987)

24. M.B. Maple, Y. Dalichaouch, J.M. Ferreira, R.R. Hake, B.W. Lee, J.J. Neumeier, M.S. Torikachvili, K.N. Yang, H. Zhou, R.P. Guertin, M.V. Kuric, $RBa_2Cu_3O_{7-\delta}$ (R = rare earth) high-T_c magnetic superconductors. Phys. B+C **148**(1–3), 155–162 (1987)

25. Y. Tokura, H. Takagi, S. Uchida, A superconducting copper oxide compound with electrons as the charge carriers. Nature **337**(6205), 345–347 (1989)

26. R.J. Cava, Structural chemistry and the local charge picture of copper oxide superconductors. Science **247**(4943), 656–662 (1990)

27. G. Uimin, J. Rossat-Mignod, Role of Cu-O chains in the charge transfer mechanism in $YBa_2Cu_3O_{6+x}$. Phys. C Supercond. **199**(3–4), 251–261 (1992)

28. S. Senoussi, Review of the critical current densities and magnetic irreversibilities in high T_c superconductors. J. Phys. III **2**(7), 1041–1257 (1992)

29. Huanbo Zhang, Hiroshi Sato, Universal relationship between T_c and the hole content in p-type cuprate superconductors. Phys. Rev. Lett. **70**(11), 1697–1699 (1993)

Chapter 5
Alternating Magnetic Shielding and Resistivity in High Critical Temperature Superconductors $(Y_{1-x}Ln_x)SrBaCu_3O_{6+\delta}$ (x = 0, 5 and 1, Ln = rare earth)

5.1 Introduction

Since its discovery in early 1986 [1], research has been directed toward the study of high critical temperature (T_c) superconducting cuprates. The alternating magnetic susceptibility ($\chi_{ac} = \chi' + i\chi''$) is very useful for characterizing high-T_c superconductors. This technique is used to study the flux dynamics of superconductors. In the real part (χ'), just below the critical temperature T_c, a strong decrease in its amplitude is a consequence of diamagnetic shielding. While in the imaginary part (χ''), the T_p peak represents the alternating losses. It is well known that high-T_c superconductivity is carried by holes in the CuO_2 planes, where the holes exhibit strong correlations among themselves. These planes are sensitive and influenced by many factors such as radius, valence, and distribution of surrounding ions and oxygen content. Moreover, the ion distribution is strongly influenced by thermal treatments applied during sample preparation. The charge transfer mechanisms in HTSCs [2] present peculiarities stipulated by the manifestation of a series of specific phenomena observed in the normal (non-superconducting) state. In general, the physical properties of these compounds are closely linked to their preparation conditions.

The compound $YBa_2Cu_3O_{6.95}$ is superconducting below its critical temperature of 92 K and characterized by double CuO_2 layers and CuO chains. The CuO_2 layers are oriented along the a-b plane and are responsible for the transport of supercurrents, while the CuO chains along the b direction are charge reservoirs for these planes [3, 4]. The majority of research on superconducting compounds is on the $ReBa_2Cu_3O_{7-\delta}$ system (with Re = Y or Lanthanide) which is justified for several reasons. On one hand, these compounds have a relatively high critical temperature $T_c \approx 90$ K above the temperature of liquid nitrogen [5, 6]. On the other hand, the electrical transport characteristics of these compounds can be varied easily by doping these compounds with substitution elements [7, 8] or by varying the oxygen content [9–11]. The preparation of samples has been carried out by several technologies with a given default structure [12, 13], which is very useful for fundamental research.

© The Author(s), under exclusive license to Springer Nature Switzerland AG 2024   97
K. Khallouq, *Exploring High-Temperature Superconductivity in the YBCO System*,
SpringerBriefs in Materials, https://doi.org/10.1007/978-3-031-66238-6_5

Among the many works on preparation, one can cite the study of the structural and superconducting properties of $R(Ba_{1-x}Sr_x)_2Cu_3O_y$ (where R is rare earth element) [14–19]. In general, for each R ion, T_c and orthorhombicity decrease with the increase in Sr concentration, and for each Sr concentration, the crystal structure and T_c depend on the size of the ionic radius of R.

We will investigate whether the effect of isovalent substitutions could modify some of the results discussed above, when Y^{3+} ($r = 0.893$ Å) is replaced by rare earth Ln^{3+} with a larger ionic radius (Eu^{3+} ($r = 0.95$ Å), Sm^{3+} ($r = 0.965$ Å), and Nd^{3+} ($r = 0.995$ Å)). In order to study the role played by the atomic planes of Yttrium and Barium and discover the factors that govern superconductivity in these compounds, we have studied the effect of thermal treatment on the structural and superconducting properties of $Y_{0.5}Ln_{0.5}SrBaCu_3O_{6+\delta}$. Indeed, we found that the influence of argon thermal treatment on these properties depended on the Ln element substituted.

5.2 Results

5.2.1 Real Part of Alternating Magnetic Susceptibility and Shielding Effect

The analysis of X-ray diffraction (XRD) diagrams of the $Y_{0.5}Ln_{0.5}SrBaCu_3O_{6+\delta}$ samples showed that monophasic compounds were obtained after the [AO] thermal treatment. The XRD diagrams of these compounds were recorded at room temperature [20, 21]. All samples allowed for the clear identification of the orthorhombic cleavage of peaks, indicating an increase in orthorhombicity and therefore in T_c, as well as the observation that a few low-intensity unidentified impurity peaks were eliminated in the [AO] samples. This indicates an improvement in the crystallographic quality of the [AO] samples. The least squares refinement of the diffraction peak positions in the 2θ XRD diagrams showed that $Y_{0.5}Ln_{0.5}SrBaCu_3O_{6+\delta}$ is orthorhombic. This is in good agreement with the values reported in the literature for YBaCuO [22]. Figure 5.1 shows the data of the alternative susceptibility of the samples as a function of temperature between 76 K and 84 K. Closed and open symbols correspond to the [O] and [AO] thermal treatments of the real and imaginary parts of the susceptibility, respectively. The effect of the [AO] thermal treatment on T_c was remarkable. The temperature at which diamagnetism sets in is taken as T_c, and it was found to depend on both the r(Ln) and the thermal treatment used in Fig. 5.1a, b. Since the same sample was used for both thermal treatments, it is possible to compare the diamagnetic response and note that the shielding current of the [AO] sample has significantly increased compared to that of the [O] sample for each r(Ln) in Fig. 5.1.

Figure 5.2 shows the variation of orthorhombicity $\epsilon = (b - a)/(b + a)$ (a, b, and c are the lattice parameters) as a function of T_c and thermal treatments of $Y_{0.5}Ln_{0.5}SrBaCu_3O_{6+\delta}$.

As the ionic radius r(Ln) decreases, orthorhombicity ϵ increases with T_c for each thermal treatment (except for T_c[AO] of Ln = Sm). For a given value of r(Ln), the

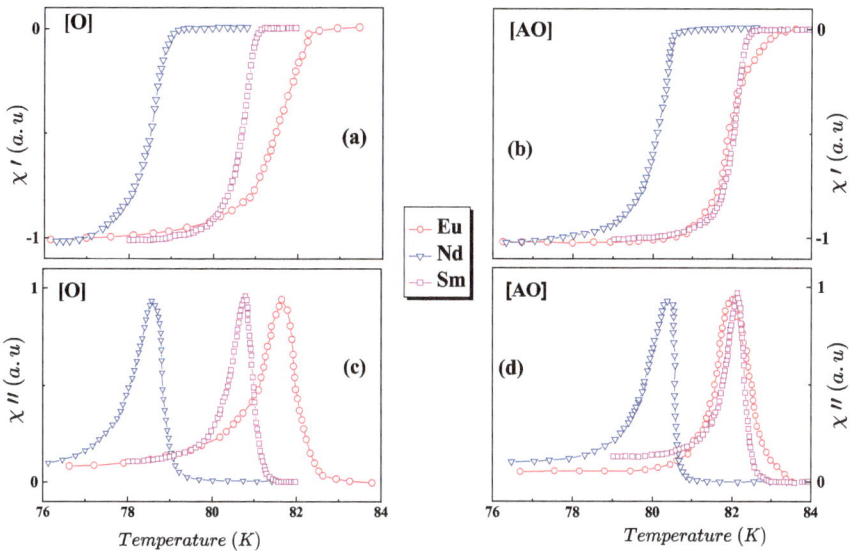

Fig. 5.1 a, b χ' and **c, d** χ'' as a function of temperature and thermal treatments of $Y_{0.5}Ln_{0.5}SrBaCu_3O_{6+\delta}$

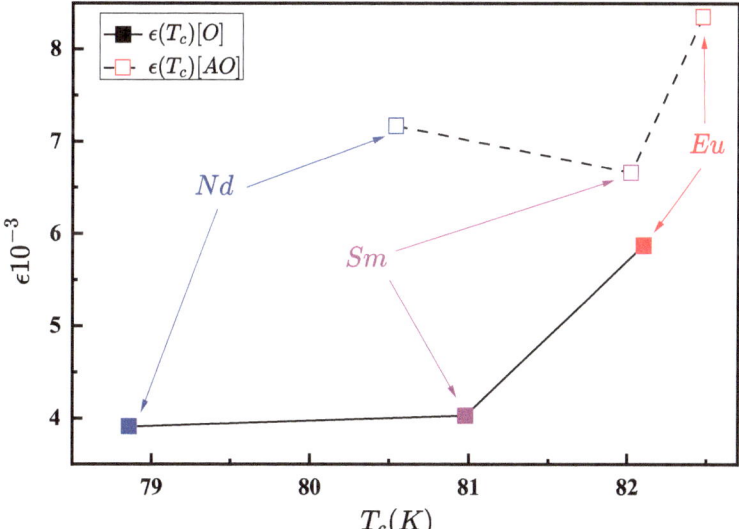

Fig. 5.2 Variation of orthorhombicity ϵ as a function of T_c and thermal treatments of $Y_{0.5}Ln_{0.5}SrBaCu_3O_{6+\delta}$

[AO] thermal treatment increases orthorhombicity ϵ and T_c, indicating an improvement in the orthorhombic structure. This improvement is due to the increase in parameter b (Fig. 4.2) by the [AO] treatment [20, 21], leading to an increase in the Cu(2)-O(2)-Cu(2) distortion angle, which decreases the distance between the

CuO charge reservoir and the CuO_2 superconducting plane, thus enhancing charge transfer and leading to an increase in T_c. The [AO] thermal treatment increases the c parameter which decreases linearly with r(Ln) for each thermal treatment [20, 21]. For a consistently low temperature, the neutron diffraction study of $RBa_2Cu_3O_{7\delta}$ (R = Y, Nd, Sm, and Eu) by Guillaume et al. [22] also showed that the crystal lattice parameters a, b, and c are influenced and evolve with the radii of the trivalent rare earth ions. For each r(Ln), the [AO] thermal treatment increases orthorhombicity and T_c. As r increases, orthorhombicity ϵ decreases with T_c.

Consider now the amplitude of the real part of the AC susceptibility in Fig. 5.1a, b, which is nothing but the shielding S [23]. S is arbitrarily set to 1 for $H_{dc} = 0$ Oe. It was measured at three different temperatures in the presence of an externally applied field H_{dc} in (Fig. 5.3). S represents the magnetic flux exclusion by the sample in an alternating dynamic mode. There was a remarkable improvement in the shielding effect in the case of the [AO] samples for all $T < T_c$ and for any applied field H for both $Y_{0.5}Eu_{0.5}SrBaCu_3O_{6+\delta}$ and $Y_{0.5}Nd_{0.5}SrBaCu_3O_{6+\delta}$ samples, but not for the $Y_{0.5}Sm_{0.5}SrBaCu_3O_{6+\delta}$ sample. For example, in the $Y_{0.5}Eu_{0.5}SrBaCu_3O_{6+\delta}$ and $Y_{0.5}Nd_{0.5}SrBaCu_3O_{6+\delta}$ samples at $T = 75$ K and $H_{dc} = 126.5$ Oe, S was improved by a factor of nearly seven times and two times, respectively, in the case of the [AO] sample compared to the [O] sample. Moreover, the decrease in S as a function of the field was much slower in the case of the [AO] sample. For example, in $Y_{0.5}Eu_{0.5}SrBaCu_3O_{6+\delta}$ at $T = 75$ K, the [AO] sample showed a decrease in S of about 10% when the field was increased from 0 to 126.5 Oe, while the [O] sample showed a decrease of nearly 70%. In the $Y_{0.5}Nd_{0.5}SrBaCu_3O_{6+\delta}$ sample, S decreased by 40% for the [AO] sample and 100% for the [O] sample. This clearly shows an improvement in the quality of the grains and intergranular coupling in the [AO] samples.

5.2.2 Resistivity

Figure 5.4 shows the DC electrical resistivity of the $LnSrBaCu_3O_{6+\delta}$ compounds. For example, the resistivity data points ρ for the [AO] sample are significantly lower than those for the [O] sample. It can be seen from these curves that the superconducting transition temperature T_c defined as the diamagnetic onset increases from 78.5 K [O] to 85 K [AO] for $EuSrBaCu_3O_{6+\delta}$, from 77.96 K [O] to 4 81.9 K [AO] for $SmSrBaCu_3O_{6+\delta}$, and from 65 K [O] to 76 K [AO] for $NdSrBaCu_3O_{6+\delta}$. Moreover, as the ionic radius size increases, the superconducting transition temperature decreases. All these results are in good agreement with those for $T_c(\chi')$. Note that for a given thermal treatment T_c (χ') is higher than T_c $(\rho = 0)$ by 2 to 3 K with $T_p(\chi'') \approx T_c(\rho = 0)$.

In the normal state, the linear part of $\rho(T)$ follows the relation $\rho = \rho_0 + \alpha T$, where ρ_0 is the residual resistivity extrapolated to $T = 0$ K and α the slope $d\rho/dT$. The [AO] treatment significantly reduced these parameters (Table 5.1). This indicates a reduction in the interaction of charges with phonons.

Fig. 5.3 Shielding effect S of $Y_{0.5}Ln_{0.5}SrBaCu_3O_{6+\delta}$ as a function of the H_{dc} field and thermal treatment at three different temperatures [20, 21]

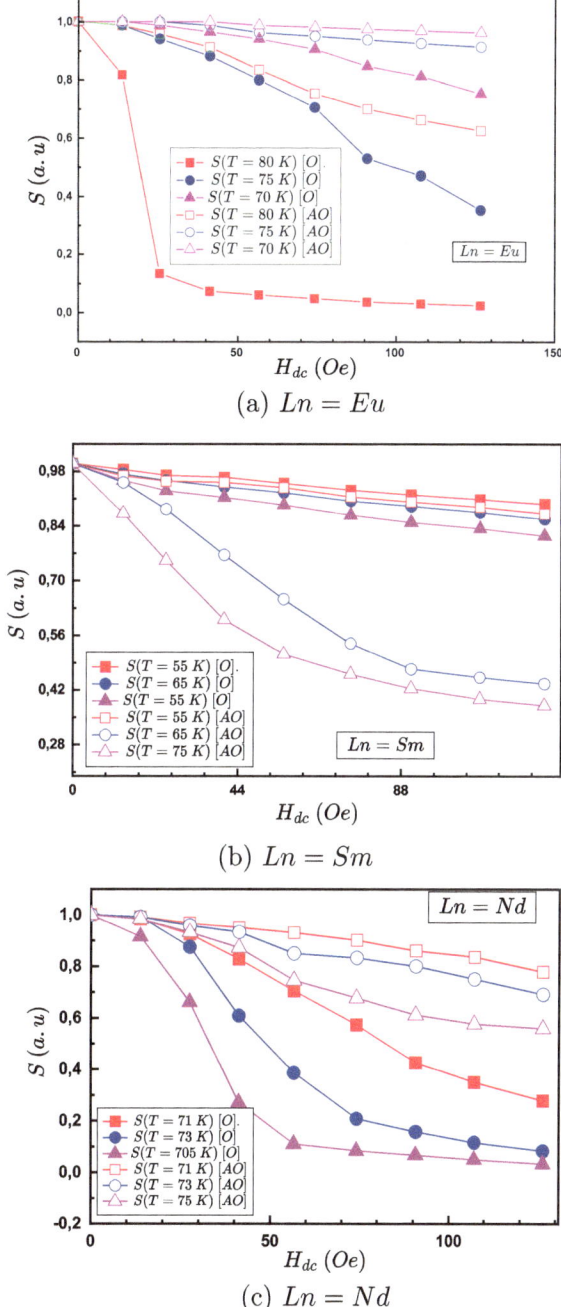

(a) $Ln = Eu$

(b) $Ln = Sm$

(c) $Ln = Nd$

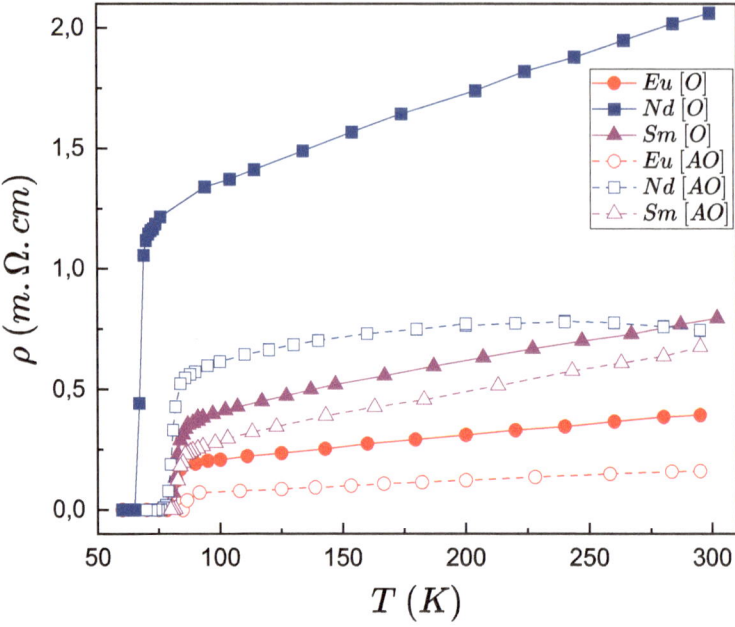

Fig. 5.4 Variation of the resistivity $\rho(T)$ of LnSrBaCu$_3$O$_{6+\delta}$ (Ln = Eu, Sm, and Nd) as a function of thermal treatment temperature [20, 21]

Table 5.1 Electrical parameters of LnSrBaCu$_3$O$_{6+\delta}$ [20, 21]

Ln	Eu		Sm		Nd	
Tr. ther.	[O]	[AO]	[O]	[AO]	[O]	[AO]
$T_c(\chi')$ (K)	81.1	87.1	79.3	84.6	68	78
$T_p(\chi'')$ (K)	80.1	86.7	79	84	67	76
$T_c(\rho = 0)$ (K)	78.5	85	77.96	81.94	65	76
ρ_0 ($\mu\Omega$cm)	120	32	242	191	1010	495
ρ_{295} ($\mu\Omega$cm)	930	161	785	455	2047	745
α ($\mu\Omega$cm/K)	0.9	0.4	1.8	0.9	3.5	1.4

5.3 Discussion

The variation in electrical and magnetic properties of Y$_{0.5}$Ln$_{0.5}$SrBaCu$_3$O$_{6+\delta}$ based on ionic radius and thermal treatment can be explained by considering the interactions between isovalent ions (Ln) and the superconducting Cu(2)O$_2$ planes. As the distance between Ln ions and CuO$_2$ increases, their interaction decreases. Thus, an increase in the size of the Ln ionic radius induces a decrease in the localization length of carriers (holes) in the $Cu(2)O_2$ plane, leading to a sudden change in electrical transport properties.

In the normal state, the [AO] thermal treatment significantly reduces the parameters of linear resistivity, indicating a decrease in the interaction of charge carriers with phonons. T_c (χ') (obtained from the real part of the alternative magnetic susceptibility measurements) and T_c ($\rho = 0$) (obtained from resistivity measurements) are in good agreement for all samples.

Figure 5.1 shows the variation in susceptibility as a function of temperature for the compounds $Y_{0.5}Ln_{0.5}SrBaCu_3O_{6+\delta}$. T_c, defined as the onset of the diamagnetic transition, is measured for both treatments as follows: 82.1 K [O], 82.5 K [AO]), (81 K [O], 82 K [AO]), and (78.9 K [O], 80.5 K [AO], respectively, for the compounds Ln = Eu, Sm, and Nd.

The highest T_c corresponds to the smallest ionic radius (Ln = Eu). For each Ln, the [AO] thermal treatment increases T_c. For all samples, T_c is also evident from the unique T_p peak $\geq T_c$ in the imaginary part of the susceptibility (Fig. 5.1). For each thermal treatment, T_c is lowest for the Nd sample (78.9 K [O], 80.5 K [AO]) and increases to a value of (82.1 K [O], 82.5 K [AO]) when moving from Nd (larger ionic radius) to Eu (smaller ionic radius).

High-temperature superconductors consist of superconducting grains coupled by an intergranular matrix (of the same chemical nature). This matrix is assimilated to Josephson junctions that connect the grains. The sample can then be considered as a network of junctions.

At low temperature, surface currents appear at the periphery of the sample. The assembly, including the grains and the surrounding matrix, is superconducting. The magnetic flux crossing the sample is shielded and weak, explaining the low values of χ' and χ''.

As the temperature increases, the coupling between the grains gradually decreases. A multitude of Josephson currents linking only a few grains and detached from each other appear.

As the temperature approaches T_p (the peak temperature of χ''), the superconducting grains become independent of each other, leading to the disappearance of Josephson currents. Each grain is traversed by a surface current.

As the temperature increases further, flux progressively penetrates the grains, and the first Abrikosov vortices appear.

χ'' reaches its maximum value when the flux reaches the center of the grain, χ'' decreases, and the sample exhibits normal metallic behavior.

Josephson junctions (grain boundaries) support Josephson currents, weakened by the applied magnetic field that penetrates the interior of the superconductor and by temperature. The shielding S decreases as Hdc and temperature increase, leading to the weakening of the links between grains by the magnetic field and temperature.

The interaction of an external magnetic field H_{dc} with our superconducting samples is influenced by several parameters, primarily: temperature, Ln ionic radius, flux pinning, magnetic field strength, and thermal treatment. All these factors are interrelated in a complex manner.

In [O] samples, as the ionic radius increases, the width of the transition ΔT_c from the superconducting state to the normal state in $\chi'(T)$ decreases. However, for a given Ln, the [AO] treatment decreases ΔT_c, due to the improvement of intergranular coupling (the matrix).

There was a remarkable improvement in the shielding effect S (amplitude of $\chi'(T)$) in the case of the $Y_{0.5}Eu_{0.5}SrBaCu_3O_{6+\delta}$ and $Y_{0.5}Nd_{0.5}SrBaCu_3O_{6+\delta}$ samples with the [AO] thermal treatment for all $T < T_c$ and for any applied field. Thus, the [AO] thermal treatment improves the quality of the grains and the intergranular coupling in the [AO] samples.

For a given r(Ln) and regardless of the thermal treatment, S decreases as temperature and H_{dc} field increase. This is the result of the deterioration of the links between grains (Josephson junctions) by temperature and magnetic field.

This effect is reversed in the $Y_{0.5}Sm_{0.5}SrBaCu_3O_{6+\delta}$ sample. The X-ray diffraction spectra of Ln = Sm in Fig. 4.1 showed no impurities regardless of the thermal treatment. For a given H_{dc} and T, the [AO] treatment decreases S, meaning that the quality of the grains and intergranular coupling have decreased.

The diffraction intensity peaks obtained by the [AO] treatment are of better quality compared to those obtained by the [O] treatment [20, 21]. These results showed the influence of the argon thermal treatment on the grain quality of the crystal structure of our $Y_{0.5}Ln_{0.5}SrBaCu_3O_{6+\delta}$ samples. The X-ray diffraction spectra showed that the impurity peaks present in the [O] samples disappeared in the [AO] samples, indicating an improvement in the crystallographic quality of the [AO] samples and an improvement in the quality of the grains and the intergranular matrix (coupling by Josephson junctions).

These results are confirmed by the fact that the difference between inter-grain and intra-grain currents disappeared because both plateaus of $\chi'(T)$ (and the two peaks in $\chi''(T)$) have merged [23] as in our case in Fig. 5.1.

In our compounds, the argon thermal treatment at 850 °C allows for the removal of oxygen and increases atomic diffusion in the structure. Thus, the departure of $(6 + \delta)$ is reduced, after argon thermal treatment, followed by annealing under oxygen flow increases the order (Y/Ln-Sr/Ba-Y/Ln) along the c-axis, which allows increasing the anionic order of O_2-in the basal plane. This leads to an increase in the number of oxygen atoms. per Cu(1)-O chain (NOC) improving hole transfer to the superconducting $Cu(2)O_2$ planes, via the apical oxygen O(1) between Cu(1) and Cu(2), and increases T_c [20, 21]. Regardless of the thermal treatment, ϵ increases with T_c as r decreases. The [AO] thermal treatment increases ϵ by $\Delta\epsilon = 3.26$ for Nd, $\Delta\epsilon = 2.634$ for Sm, and $\Delta\epsilon = 2.483$ for Eu (Fig. 5.2).

For each thermal treatment, as the ionic radius r(Ln) increases, the crystalline parameter b is almost constant [20, 21], but parameter a increased from Eu to Sm then decreased for Nd, decreasing c. This indicates an increase in the number of oxygen atoms per chain (NOC) along the a-axis leading to a decrease in ϵ (T_c) with an orthorhombic symmetry.

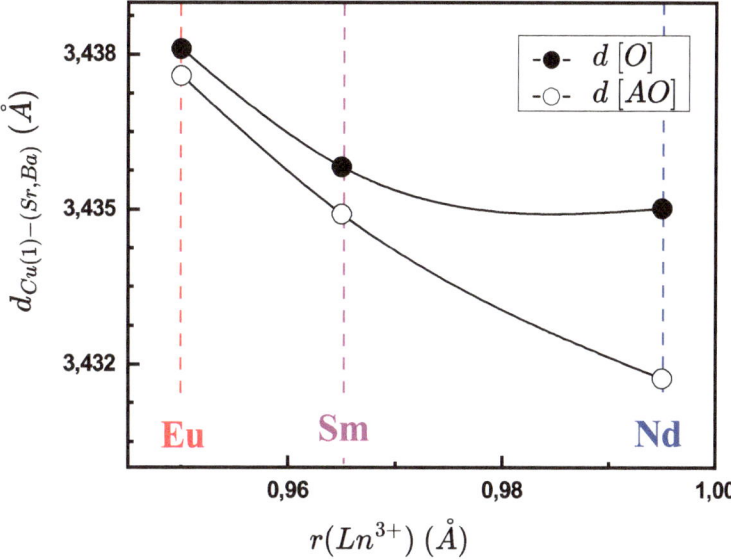

Fig. 5.5 Interatomic distance $d_{Cu(1)-(Sr/Ba)}$ as a function of the ionic radius r_{Ln}^{3+} and thermal treatment of $Y_{0.5}Ln_{0.5}SrBaCu_3O_{6+\delta}$ [20, 21]

Whereas for each r, the [AO] treatment increases orthorhombicity ϵ, T_c, and decreases the distance $d_{Cu(1)-(Sr,Ba)}$ (Fig. 5.5). This improves charge transfer from chains along the b-axis to the superconducting $Cu(2)O_2$ planes via the apical oxygen O(1) along the c-axis. Moreover, for each r, the [AO] thermal treatment decreases a and increases b. This increases the value of (NOC) along the b-axis, decreases the cationic disorder of Y/Ln on the Ba/Sr site along the c-axis, and increases the anionic order in the basal plane, leading to an increase in the number of holes $\rho(r)$ in the superconducting $Cu(2)O_2$ planes (deduced from the sub-saturation zone of the universal relation $T_c / T_c^{max} (\rho)$ [24]) and T_c (Fig. 5.6). The decrease in the distance $d_{Cu(1)-(Sr,Ba)}$ (Fig. 5.5) can be considered as an argument in the increase in the formation of Cooper pairs, ρ, and T_c.

Both arguments (cationic and anionic disorders) are justified here by the four remarkable correlations observed between $T_c(r)$ and $V(r)$ (Fig. 5.7); between $T_c(r)$ and $T_c(d_{Cu(1)-(Sr,Ba)})$ (Fig. 5.8); between the number of holes $\rho(r)$ and $T_c(r)$ (Fig. 5.6); and, on the other hand, between $\Delta T_c(r) = T_c[AO] - T_c[O]$ and $\Delta\epsilon(r)$ in Fig. 5.9. Thus, the structural, electrical, and magnetic properties in these superconductors are correlated with the effect of argon thermal treatment.

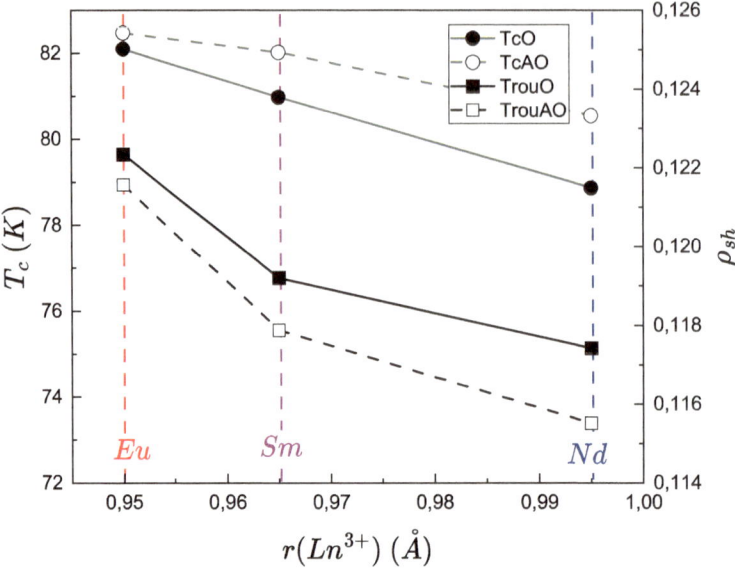

Fig. 5.6 Correlation between ρ and T_c, as a function of the ionic radius r(Ln^{3+}) and thermal treatment of $Y_{0.5}Ln_{0.5}SrBaCu_3O_{6+\delta}$ [20, 21]

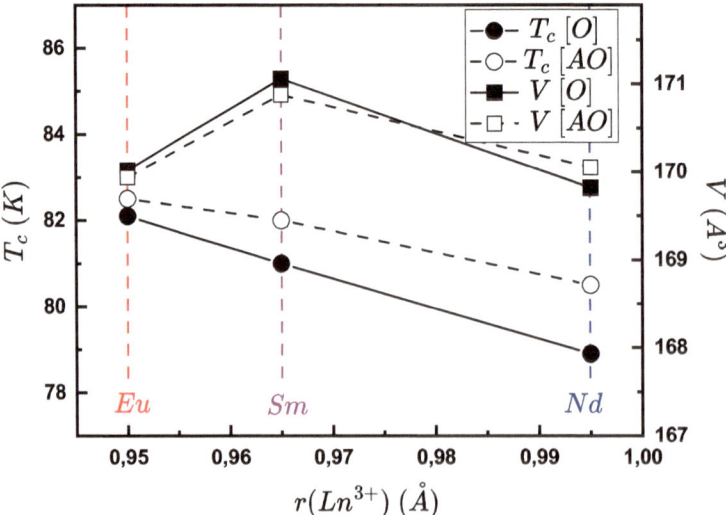

Fig. 5.7 Correlation between volume V(r) and T_c (r) as a function of thermal treatment of $Y_{0.5}Ln_{0.5}SrBaCu_3O_{6+\delta}$ [20, 21]

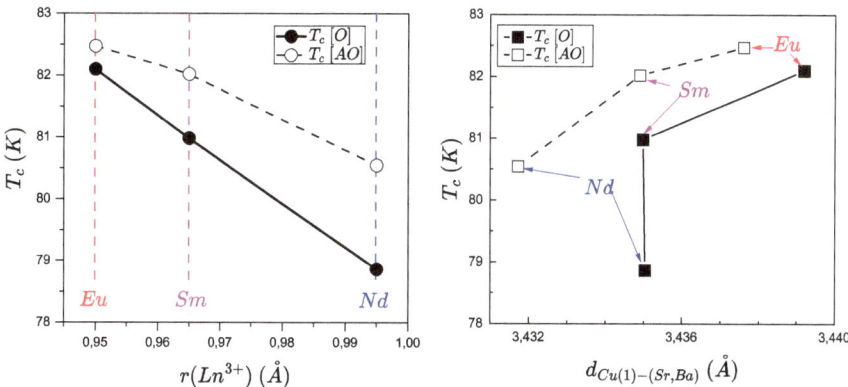

Fig. 5.8 Correlation between T_c ($d_{Cu(1)-(Sr,Ba)}$) and T_c (r) as a function of thermal treatment of $Y_{0.5}Ln_{0.5}SrBaCu_3O_{6+\delta}$ [20, 21]

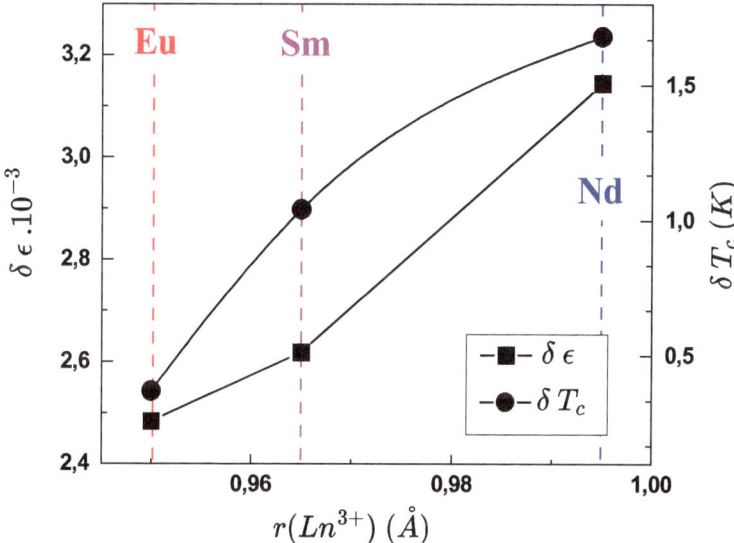

Fig. 5.9 Correlation between $\Delta T_c = T_c$ [AO] $- T_c$[O] and $\Delta\epsilon = \epsilon$ [AO] $- \epsilon$[O] as a function of the thermal treatment of $Y_{0.5}Ln_{0.5}SrBaCu_3O_{6+\delta}$ [20, 21]

5.4 Conclusion

X-ray diffraction spectra, AC magnetic susceptibility, and resistivity have been measured in the $Y_{0.5}Ln_{0.5}SrBaCu_3O_{6+\delta}$ system as a function of r(Ln^{3+}) (where Ln = rare earth = Eu, Sm, and Nd) and thermal treatment. The isovalent substitutions have induced several modifications in the structural, electrical, and magnetic properties. Our data explicitly show that in the [AO] samples, the remarkable increase in the shielding effect is explained by the improvement in the quality of intergranular coupling resulting from the enhancement in the crystallographic quality of these samples. The increase in the number of holes and oxygen atoms localized respectively in the superconducting $Cu(2)O_2$ planes and the CuO chains in the basal plane is in agreement with the charge transfer model. These results stem from an interaction between the cationic disorder of Ln^{3+} on the Sr/Ba site along the c-axis and the anionic disorder of oxygen in the basal plane.

A combination of several factors such as the decrease in $d_{Cu(1)-(Sr,Ba)}$, the increase in cationic order and the order of oxygen in the chains, ρ, and the phase purity of the [AO] samples could explain the observed results. We believe these results will be useful for testing or improving certain theoretical models on the electronic structure and atomic disorder.

References

1. P.H. Hor, L. Gao, R.L. Meng, Z.J. Huang, Y.Q. Wang, K. Forster, J. Vassilious, C.W. Chu, M.K. Wu, J.R. Ashburn, C.J. Torng, High-pressure study of the new Y-Ba-Cu-O superconducting compound system. Phys. Rev. Lett. **58**(9), 911–912 (1987)
2. J. Ashkenazi, A theory for the high-t_c cuprates: anomalous normal-state and spectroscopic properties, phase diagram, and pairing. J. Supercond. Novel Magn. **24**(4), 1281–1308 (2010)
3. Y. Tokura, H. Takagi, S. Uchida, A superconducting copper oxide compound with electrons as the charge carriers. Nature **337**(6205), 345–347 (1989)
4. R.J. Cava, Structural chemistry and the local charge picture of copper oxide superconductors. Science **247**(4943), 656–662 (1990)
5. B. Raveau, C. Michel, M. Hervieu, D. Groult, *Crystal Chemistry of High-T_c Superconducting Copper Oxides* (Springer, Berlin Heidelberg, 1991)
6. R.V. Vovk, N.R. Vovk, G.Y. Khadzhai, I.L. Goulatis, A. Chroneos, Effect of high pressure on the electrical resistivity of optimally doped $YBa_2Cu_3O_{7-\delta}$ single crystals with unidirectional planar defects. Phys. B: Condens. Matter **422**, 33–35 (2013)
7. R.V. Vovk, N.R. Vovk, G.Y. Khadzhai, I.L. Goulatis, A. Chroneos, Effect of praseodymium on the electrical resistance of $YBa_2Cu_3O_{7-\delta}$ single crystals. Solid State Commun. **190**, 18–22 (2014)
8. H.A. Borges, M.A. Continentino, Pressure study of the paraconductivity of high T_c superconductors. Solid State Commun. **80**(3), 197–199 (1991)
9. M.K. Wu, J.R. Ashburn, C.J. Torng, P.H. Hor, R.L. Meng, L. Gao, Z.J. Huang, Y.Q. Wang, C.W. Chu, Superconductivity at 93 K in a new mixed-phase Y-Ba-Cu-O compound system at ambient pressure. Phys. Rev. Lett. **58**(9), 908–910 (1987)
10. R.J. Cava, B. Batlogg, C.H. Chen, E.A. Rietman, S.M. Zahurak, D. Werder, Oxygen stoichiometry, superconductivity and normal-state properties of $YBa_2Cu_3O_{7-\delta}$. Nature **329**(6138), 423–425 (1987)
11. R.V. Vovk, N.R. Vovk, O.V. Dobrovolskiy, Effect of structural relaxation on the in-plane electrical resistance of oxygen-underdoped $REBa_2Cu_3O_{7-\delta}$ ((Re = Y, Ho) single crystals. J. Low Temp. Phys. **175**(3–4), 614–630 (2014)
12. P. Schleger, W.N. Hardy, B.X. Yang, Thermodynamics of oxygen in $Y_1Ba_2Cu_3O_x$ between 450° and 650°. Phys. C: Supercond. **176**(1–3), 261–273 (1991)

13. R.V. Vovk, Z.F. Nazyrov, I.L. Goulatis, A. Chroneos, Localization effect and pseudogap in praseodymium doped $Y_{1-z}Pr_zBa_2Cu_3O_{7-\delta}$ single crystals. Modern Phys. Lett. B **26**(25), 1250163 (2012)
14. T. Wuernisha, Y. Takahashi, K. Takase, Y. Takano, K. Sekizawa, Effects of sr substitution for ba on the crystal structure, oxygen content and electrical properties of $Nd(Ba_{1-x}Sr_x)_2Cu_3O_{7-\delta}$. J. Alloys Comp. **377**(1–2), 216–220 (2004). (September)
15. R.A. Gunasekaran, B. Hellebrand, P.L. Steger, Crystal structure, oxygen stoichiometry and superconducting properties of $GdBa_{2-x}Sr_xCu_3O_{7-\delta}$ ($0.0 \leq x \leq 1.6$). Phys. C: Supercond. **270**(1–2), 25–34 (1996)
16. V.P.S. Awana, C.A. Cardoso, O.F. de Lima, S.K. Malik, W.B. Yelon, R. Prasad, A. Gupta, A. Sedky, A.V. Narlikar, Rare earth ionic size dependence of T_c in $RBaSrCu_3O_7$ (R = Y, Dy, Nd, and La) series. Phys. C: Supercond. 341–348, 627–628 (2000)
17. X.Z. Wang, B. Hellebrand, D. Bäuerle, Crystal structure and superconductivity in REBaSrCu$_3O_x$. Phys. C: Supercond. **200**(1–2), 12–16 (1992). (September)
18. P. Karen, H. Fjellvåg, A. Kjekshus, A.F. Andresen, Chemical pressure and other effects of strontium substitution in $YBa_2Cu_3O_{9-\delta}$. J. Solid State Chem. **92**(1), 57–67 (1991). (May)
19. Y.G. Zhao, S.Y. Xiong, Y.P. Li, B. Zhang, S.S. Fan, B. Yin, J.W. Li, S.Q. Guo, W.H. Tang, G.H. Rao, D.J. Dong, B.S. Cao, B.L. Gu, Structural, transport, and magnetic properties of $PrBa_{2-x}Sr_xCu_3O_{7-\delta}$. Phys. Rev. B **56**(14), 9153–9157 (1997). (October)
20. K. Khallouq, A. Nafidi, A. Aboulkassim, E. Youssef El Yakoubi, E. Es-Salhi, Effects of isovalent substitutions and heat treatment on structural and superconducting properties of high-critical temperature superconductors. Mater. Today: Proc. **22**, 140–145 (2020)
21. K. Khallouq, A. Nafidi, A. Aboulkassim, M. Bellioua, E. Youssef El Yakoubi, A. Khalal, Effects of isovalents substitutions and argon heat treatment on the structural and superconducting properties of $Y_{0.5}Ln_{0.5}SrBaCu_3O_{6+z}$. Int. J. Eng. Res. Technol. **5**(10), 179 (2016). Total Downloads: 179
22. M. Guillaume, P. Allenspach, W. Henggeler, J. Mesot, B. Roessli, U. Staub, P. Fischer, A. Furrer, V. Trounov, A systematic low-temperature neutron diffraction study of the $RBa_2Cu_3O_x$ (R = yttrium and rare earths; x = 6 and 7) compounds. J. Phys.: Conden. Matt. **6**(39), 7963–7976 (1994)
23. R.A. Hein, T.L. Francavilla, D.H. Liebenberg (eds.), *Magnetic Susceptibility of Superconductors and Other Spin Systems* (Springer, US, 1991)
24. H. Zhang, H. Sato, Universal relationship between T_c and the hole content in p-type cuprate superconductors. Phys. Rev. Lett. **70**(11), 1697–1699 (1993). (March)

General Conclusion

In concluding a comprehensive exploration of superconductivity, this book has traversed from foundational theories and historical milestones through the nuanced phenomena of high critical temperature superconductors, and culminating in detailed studies of the YBCO system and the impact of isovalent substitutions and thermal treatments.

Starting with the serendipitous discovery in 1911, superconductivity has unfolded as a field of profound theoretical and practical significance. The journey from early applications to the development of phenomenological and microscopic theories provided a solid foundation for understanding superconducting materials. The London theory, Ginzburg–Landau theory, and the groundbreaking BCS theory illuminated the fundamental mechanisms underlying superconductivity, highlighting the roles of penetration depth, coherence length, and the formation of Cooper pairs.

The narrative progressed to the era of high critical temperature superconductors, marking a revolutionary phase in the quest for superconductivity at more accessible temperatures. The discovery of cuprates and iron-based superconductors expanded the landscape, challenging existing theories and inspiring new lines of inquiry into the electronic structure and atomic disorder of these complex materials.

In-depth examinations of the YBCO system revealed the intricate relationships between crystal structure, oxygen stoichiometry, charge transfer, and superconducting properties. Experimental techniques, from X-ray diffraction to resistivity measurements, facilitated a nuanced understanding of the effects of substitutions and thermal treatments on these materials.

Isovalent substitutions along with argon heat treatments have notably impacted the structural and superconducting characteristics of $Y_{0.5}Ln_{0.5}SrBaCu_3O_{6+\delta}$ compounds. Such alterations highlight the intricate relationship among lattice parameters, oxygen levels, and carrier concentration in shaping the superconducting properties.

The book concludes by acknowledging that superconductivity, especially in high critical temperature materials, remains a fertile ground for research. The intricate balance between theoretical models and experimental findings continues to drive the

K. Khallouq, *Exploring High-Temperature Superconductivity in the YBCO System*,
SpringerBriefs in Materials, https://doi.org/10.1007/978-3-031-66238-6

field forward, offering potential pathways to new technologies and deeper understanding of quantum phenomena. The journey from the macroscopic manifestations of superconductivity to its microscopic origins and back to its technological applications illustrates the cyclical nature of scientific discovery, where each answer unfolds new questions, and every theory finds its ultimate test in the crucible of experiment. This exploration not only enriches our comprehension of superconductivity but also highlights the enduring fascination and challenge of uncovering the mysteries of the quantum world.